CULINARY REACTIONS

CULINARY REACTIONS

THE EVERYDAY CHEMISTRY OF COOKING

SIMON QUELLEN FIELD

Library of Congress Cataloging-in-Publication Data

Field, Simon (Simon Quellen)

Culinary reactions : the everyday chemistry of cooking / Simon Quellen Field.

p. cm.

Includes index.

Summary: "When you're cooking, you're a chemist! Every time you follow or modify a recipe, you are experimenting with acids and bases, emulsions and suspensions, gels and foams. In your kitchen you denature proteins, crystallize compounds, react enzymes with substrates, and nurture desired microbial life while suppressing harmful microbes. And unlike in a laboratory, you can eat your experiments to verify your hypotheses. In CULINARY REACTIONS, author Simon Field explores the chemistry behind the recipes you follow every day. How does altering the ratio of flour, sugar, yeast, salt, butter, and water affect how high bread rises? Why is whipped cream made with nitrous oxide rather than the more common carbon dioxide? And why does Hollandaise sauce call for "clarified" butter? This easy-to-follow primer even includes recipes to demonstrate the concepts being discussed, including Whipped Creamsicle Topping (a foam), Cherry Dream Cheese (a protein gel), and Lemonade with Chameleon Eggs (an acid indicator). It even shows you how to extract DNA from a Halloween pumpkin. You'll never look at your graduated cylinders, Bunsen burners, and beakers—er, measuring cups, stovetop burners, and mixing bowls—the same way again"— Provided by publisher.

ISBN 978-1-56976-706-1 (pbk.)

1. Food—Analysis. 2. Cooking. I. Title.

TX545.F46 2012

664'.07—dc23

2011029366

Cover design: John Yates at Stealworks.com

Cover photograph: Sabine Scheckel/Photodisc/Getty Images

Interior design: Scott Rattray

Published by Chicago Review Press, Incorporated

814 North Franklin Street

Chicago, Illinois 60610

ISBN 978-1-56976-706-1

Printed in the United States of America

5 4 3 2 1

To Kathleen, my favorite chef

CONTENTS

INTRODUCTION

Your mother was a chemist. In the kitchen, she experimented with acids and bases, emulsions, suspensions, gels, and foams. She denatured proteins, crystallized compounds, reacted enzymes with substrates, and nurtured desired microbial life while suppressing harmful microbes. In other words, she cooked your dinner.

Cooking is often about combining ingredients to create something completely different. It involves many chemical and physical changes to the food that the cook carefully controls in order to produce the desired result. This book is about those changes. Understanding them might help make you a better cook, but my aim here is mostly to have fun.

You can learn a lot of science in the kitchen. But just looking at food in a different way can be fun and enlightening. How many of your favorite foods are foams? Bread, cake, whipped cream, marshmallows, ice cream, and meringue—all would be quite different if they didn't have bubbles of gas in them. What makes some foods foam and others not? What happens when you heat a foam? What is actually going on in the bread that changes it from a sticky, runny dough or batter into a structural element that holds a sandwich together?

Knowing how things work also helps when you want to make changes to a recipe. What would you have to do if you wanted a harder cookie, or a softer one? What went wrong when you tried to make fudge but got a hard lump of rock in the pan instead? If you don't want to use an ingredient that's less than healthy or that you are allergic to, what should you replace it with? What other changes will you have to make?

A LITTLE WHILE BACK I made a big batch of ice cream for a group of Nobel Prize winners and other brilliant scientists at a scientific convention. I brought along a huge 160-liter Dewar flask of liquid nitrogen, and we made ice cream. At –321°F (–196°C), the liquid quickly cooled the ingredients to the right temperature. But at the same time, the nitrogen boiled vigorously, making a foam of nitrogen gas (basically air without the oxygen) to whip up the ice cream. Instead of a rock-hard chunk of ice, we got something closer to soft-serve—wonderfully smooth, the ice crystals so tiny the tongue mistook them for cream.

It is in that spirit that these pages will continue. Let's have fun. Let's play with our food.

Simon
LD10

1

MEASURING AND WEIGHING

In science, and especially in chemistry, careful weighing and measuring are important for reproducible results. If someone cannot reproduce your results, there is little point in doing the experiment.

For people to reproduce a culinary masterpiece, it is important to carefully weigh and measure according to a recipe. But when you're just cooking up some breakfast, it is more important to know *why* the ingredients are used, and why certain processes are followed. With this knowledge, you create and adjust the food on the fly, substituting some ingredients you have for some you don't, or use up things from the back of the refrigerator before they go bad.

Variations in Recipes

You can get a feel for how important measuring is by comparing recipes. Suppose you look at 10 recipes for homemade cupcakes and compare the ratios of flour and sugar in them:

	Flour	Sugar	Ratio
	1.5	1	150.00%
	2.75	1.5	183.33%
	2	1.5	133.33%
	1.5	1	150.00%
	2	2	100.00%
	3	2	150.00%
	2	2	100.00%
	2.5	1	250.00%
	3	2	150.00%
	2.5	2	125.00%
Mean			**149.17%**
Standard Deviation			**43.47%**

The average cupcake has one and a half times as much flour as sugar. But some cupcakes have equal amounts, and some have two and a half times as much flour as sugar. The high standard deviation means that there is a lot of variation among simple cupcake recipes. A good cook can feel free to vary the amount of sugar in the recipe for taste or to compensate for what will accompany the cake, such as icing or bits of fruit in the batter.

Why Sifted Flour?

Some recipes list the ingredients by weight instead of volume. Some cooks swear by weighing everything, to get consistent results. When consistency of results is important, by all means, measure carefully. But when a little variation and creativity are called for, or when you are changing parts of the recipe for whatever reason, judgment and knowledge are more important.

Recipes once called for sifting flour. Flour was something that often had lumps, bits of millstone, or insects in it, so sifting was important. Other reasons have been suggested for sifting, such as aeration, or mixing dry ingredients, but a whisk in a bowl can accomplish both these tasks. The bother of sifting would not be worth it if either of these were the main reason.

So why sift? When ingredients are not weighed, the difference between a cup of flour and a cup of sifted flour can be significant. But a knowledgeable cook can use a bit less flour and avoid the time and mess of sifting.

It is interesting to look at recipes that are very careful to weigh out all of the ingredients yet then call for three eggs, without specifying the weight of the eggs. Eggs vary in weight, but most recipes don't specify the size of the eggs as small, medium, large, extra large, or jumbo. The reason is that it really doesn't matter too much. Whatever the size, the recipe is going to come out just fine. There is a lot of room for variation, and consistent results are usually not as important to the eater as they are to the creator of the recipe, who wants to protect his or her reputation for being reliable.

The best recipes will tell you what to look for in the processing of the food. Instead of giving a precise baking time, a cake will be tested for doneness with a toothpick or the press of a finger. In candy making, the initial amounts of sugar and water are not that important when you are cooking the mixture to a certain temperature or to "hard ball" stage, both of which are measures that tell the cook exactly what the ratios are during cooking.

Density and Good Eggs

In making wine or beer, the density of the mixture, measured by floating a little scale (called a *hydrometer*) in the water, tells how much sugar, alcohol, and water are in the mix at any given time. A density test can also tell you how fresh your eggs are. Place an egg in water, then dissolve measured amounts of salt into the water until the egg floats. A bad egg will float right away.

You may have noticed at a party that some cans of soda in a tub of ice water float, while others sink. This is caused by density; sodas with sugar in them are at the bottom and the diet sodas are at the top. As an interesting experiment, place a diet soda can in a glass container large enough for it to float, then place a small plastic cup on top. Slowly fill the cup with sugar until the can sinks. You might be amazed at how much sugar it takes to sink the can. There is at least that much sugar in the sodas that sink, but probably more.

Another place where density comes into play in the kitchen is in making hard-boiled eggs. The yolk of an egg contains fats and oils and is thus less dense than the white of the egg. This means that if left to itself, the yolk inside will float to the top of the egg and thus be off-center when the egg is cut in half for deviled eggs or sliced into a salad.

To keep the yolk centered, the eggs must be turned frequently while being cooked, keeping the yolk away from the shell. Since the white of the egg cooks on the outside first (where it is closer to the boiling water), the yolk that is turned often will not be able to get past the hardening white and will end up centered.

Calorie Estimation

Some things are easy to measure. Not all cooks have kitchen scales, so many recipes (especially in the United States) call for easy volume measurements. But some things you might care about, such as how many calories are in the food you are making, might at first seem hard to measure at home.

But with a little thought, estimating calories isn't that difficult.

As a general rule, proteins and carbohydrates have about 4 calories per gram, while fat has about 9. You can separate the ingredients by whether they are fats or not, weigh them, and then multiply. Or you can estimate by eye what percentage of the recipe is fats, and pick a number between 4 and 9 that matches the estimate. A little adjustment for water content, and you have a good guess at the number of calories in the food.

A Hostess Twinkie says on the label that it has 4.5 grams of fat (40.5 calories) and 27 grams of carbohydrates (108 calories) for a total of 148.5 calories. One Twinkie weighs 43 grams, and the label says it has 150 calories, so about 3.5 calories per gram.

Take a look at some popular foods:

- Beef jerky: 116 calories in 28 grams, or 4 calories per gram
- Pork sausage: 95 calories in 28 grams, or 3.4 calories per gram
- Air-popped popcorn: 31 calories in 8 grams, or 3.8 calories per gram
- Butter: 70 calories in 10 grams, or 7 calories per gram
- Bacon: 50 calories in 12 grams, or 4 calories per gram
- Buttercream frosting: 100 calories in 26 grams, or 3.8 calories per gram
- Enriched flour: 455 calories in 125 grams, or 3.6 calories per gram
- Whole wheat bread: 70 calories in 28 grams, or 2.5 calories per gram
- A steak: about 2 calories per gram

What you see from the examples above is that until you get to something like pure butter, most processed foods have between 3½ to 4 calories per gram, about the same as pure sugar.

Celery has 0.16 calories per gram, an apple has 0.5 calories per gram, and a carrot has 0.4 calories per gram. These foods are mostly water. So eat fruits and vegetables to fill yourself up if you are watching your calories.

Steaks, chicken, pork chops—even those have fewer calories per gram than popcorn or bread. But within about a factor of two, you can simply weigh the food and figure 1,300 to 1,800 calories per pound. Put your whole meal on a plate and weigh it. If you don't like what the bathroom scale says the next morning, put less on your plate today.

Of course, counting calories to control your weight assumes that your weight is simply a matter of balancing the number of calories you eat with the number of calories you burn. But your body already has mechanisms for doing that balancing. If you starve yourself, your body will stop burning as many calories. If you eat too much, your body will burn more. This is controlled by hormones in your body, the main one being insulin.

Insulin tells the fat cells to take in sugar from the blood. When there is too much sugar in the blood, extra insulin is produced to remove it, and thus extra fat is stored. Foods with high insulin indexes (foods that cause more insulin to be produced than other foods do) can upset the balance that keeps your calorie inputs and outputs matched. This is why low-carbohydrate diets seem to be effective in controlling weight. They prevent excess insulin from being produced and thus prevent extra fat from being stored.

There are many complex interactions in the body that affect the balance that controls fat production. Some are genetic, some are behavioral, some are environmental, and some are caused by infections or disease. Planning effective weight control for an individual will necessarily be an individual exercise, and one diet plan will not work for everyone. But it is important to understand that simply cutting calories or getting more exercise is not the whole story.

❖ 2 ❖

FOAMS

Foams are fun. Marshmallows, meringues, cakes, whipped cream, cookies, ice cream—all of these are foams.

Foams are formed by several different processes. In many foams, such as whipped cream and beaten egg whites, an interesting thing happens at the interface between water and air. In both of these foams, proteins in the foam are first *denatured*, which, as the name implies, means that they are changed from their natural state.

Proteins are made up of building blocks called *amino acids*. Some of these building blocks are attracted to water but avoid oils and fats. Others are attracted to oils and fats but are repelled by water. In the natural state of the protein, the water-loving parts are on the outside of the protein, next to the water, and the water-avoiding parts are tucked inside, away from the water.

Proteins are big molecules, formed of strands and sheets of amino acids, all tangled up into a shape that is important for their natural function. When we beat the cream or the egg whites, the protein unfolds, like a carefully folded origami animal would if you beat it hard with a whisk.

As the protein unfolds, it encounters oils and fats in the cream, as well as air. The water-loving parts of the protein still stay in the water. The water-avoiding parts unfold so they can stick into the fats or into the air, to avoid the water. Eventually, the air bubbles become smaller and smaller as they are beaten, and they become surrounded by a film made of protein, to which some water is still attached. The proteins can now link together to form a tough film that holds the bubbles in shape and prevents them from merging together again.

In whipped egg whites, you get bubbles with a protein film. The water-loving parts stick into the water, and the water-avoiding parts stick into the air bubble.

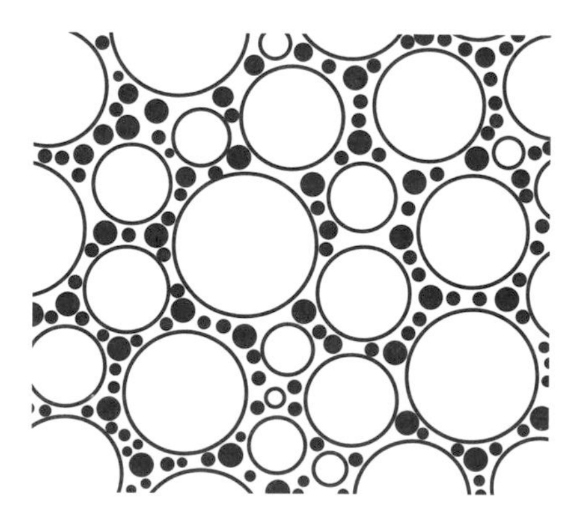

In whipped cream, you get big bubbles of air surrounded by a film of protein, surrounded by tiny globules of fat stuck to the fat-loving parts of the protein, connected to another film of protein that forms the wall of the next bubble. In between the bubbles of air and the globules of fat, the water-loving parts of the proteins extend into the water.

Egg Foams

You can make an egg white foam more stable by increasing the number of places where the proteins bond together. Beating the egg whites in a copper bowl causes the amino acids that have sulfur in them to bond together where the sulfur atoms are. Linking two sulfur atoms in this way forms a *disulfide bridge*, a very strong chemical bond that helps keep the protein stuck in the new position.

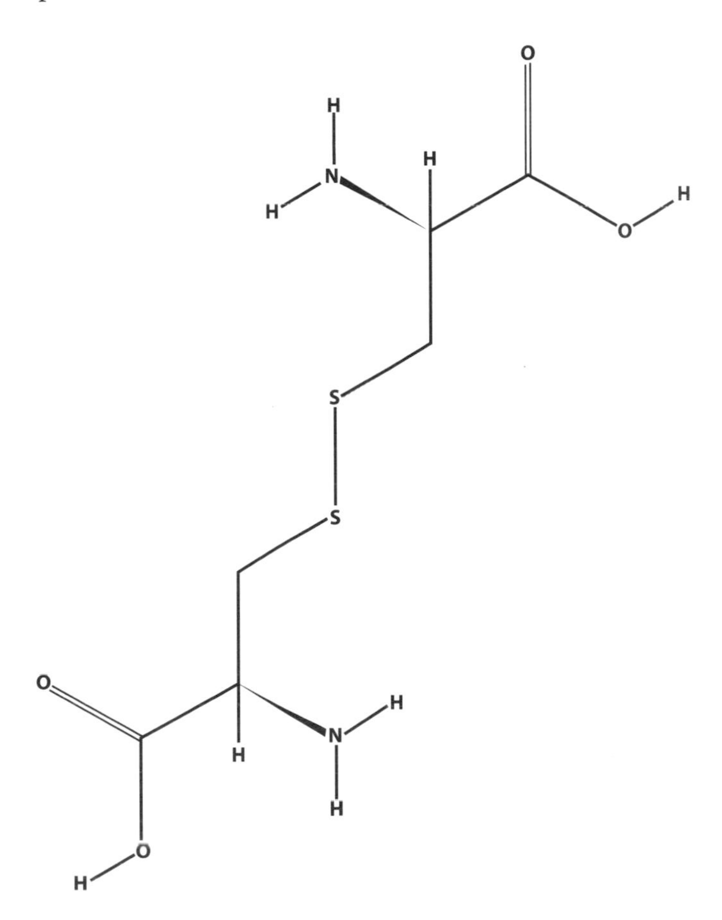

Cystine is an example of a disulfide bridge

Adding an acid such as lemon juice or cream of tartar can also help form more bonds between the proteins and stabilize the foam, because the acid unravels the protein a bit, allowing the proteins to tangle and bond together.

Chemistry Lesson

How to Read Structural Formulas

Chemists use some simplifying conventions to show how a molecule is shaped without cluttering up the picture.

Since carbon atoms are so common, they are not labeled with a "C." Instead, they are assumed to be anywhere on a formula where two lines join. And since hydrogens attached to carbons are also very common, and carbon always has four bonds, any place on a formula where fewer than four lines join, it is assumed that hydrogens fill the carbon's remaining bonds, and so they are not labeled.

If a line looks more like a dark wedge, it means that portion of the molecule comes out of the page toward the reader. If the wedge is lighter in color, it goes into the page, away from the reader.

Two lines mean a double bond, and three means a triple bond.

Fat Foams

You probably know that whipped cream forms a foam but whipped milk does not (unless it is heated with steam). The reason lies in the nature of the proteins in milk and cream, and the nature of butterfat. But mostly it lies in the amount of solid fat compared to the amount of water.

Butterfat is a liquid at body temperature—anything above about 90°F (32°C)—but it solidifies when chilled. This is why butter melts in your mouth. To whip cream, you need chilled, solid butterfat. As you beat the cream, it forms bubbles and the proteins denature, with some parts staying in the water and some parts staying in the fat, until you end up with a film of solid fat and protein that traps the air inside, with the water in between the bubbles.

If you beat the cream too much, you can turn the whole thing inside out, with the water trapped inside films of fat and protein, and the air gets out. This is butter. Where cream was tiny bits of fat in liquid water, butter is tiny drops of water in solid fat. One is a liquid and the other is a solid, but both are made of the same stuff.

To keep whipped cream stable, you need to keep the temperature low enough that the fat stays quite solid. You can also stabilize it by adding more protein, such as gelatin or some vegetable gums. Both help to link the proteins together and hold the fat in place.

If cream does not contain at least 30 percent fat, it will be difficult to whip. Most whipping cream is about 36 percent fat.

Reduced fat whipping creams need the help of stabilizers. Most common are cellulose-based ingredients called hydrocolloids, or food gums.

As you whip cream, it gradually becomes stiffer. Maximum stiffness happens when the cream just starts to become butter. It will be slightly yellow in appearance, and the volume will have dropped a bit. The stiffness comes from the firm butterfat that has formed larger and larger particles on its way to becoming butter.

If your recipe uses whipped cream as a structural element, such as in cake icing or rosettes on a cream pie, you will want a nice stiff cream. For toppings on strawberry shortcake or other desserts, stopping the whipping when the foam is at peak volume will make it stretch further.

If you want to make a foam out of milk, you must use steam, as in a cappuccino machine. The steam denatures the proteins and links them together and at the same time incorporates air into the foam. When the steam cools, it becomes water again. The foam is full of air, not steam.

Gluten Foams

Wheat flour contains a protein called *gluten*, which is formed when enzymes in the flour react with precursor proteins as water is added. Gluten is gluey, and as you mix the batter or knead the dough, the little bits of gluten that form stick together and form rubberlike sheets.

Stirring and folding incorporate air to form little bubbles in the sheets of protein. Yeast or other leavening agents add gas inside the bubbles and make them expand. Heating the dough further changes the protein, denaturing it into a solid.

A Bread Recipe

Basic bread is fairly simple. You need some flour, some water, some yeast, and optionally some sugar or honey, salt, and/or oil, butter, or some other fat.

What do those ingredients do, and how much of each do you need? The flour provides the gluten precursors, starch, flavor, and bulk of the bread. Water is necessary to make the gluten and allow the yeast to multiply and produce carbon dioxide gas. The yeast is there to make the carbon dioxide gas so you get a foam instead of a brick.

All the other ingredients are optional. The salt is not there just as a seasoning; it's there to slow down the yeast. (There really isn't a lot of it in most breads.) If the yeast produces too much gas too fast, faster than the gluten forms, the gas will simply escape as the bubbles pop. But many recipes omit the salt. Some of the gas will escape, but these recipes usually call for the size of the bread to double, which will eventually happen with or without the salt.

Sugar or honey is often added to feed the yeast. But the yeast will find enough food in the flour without it. It will just grow a little more slowly, which (as we saw with adding salt) can be a good thing. But if you are making a lot of bread, and start with a small amount of yeast, you can grow the yeast you need in a little sugar water. The amount of sugar or honey is generally so small that it makes little difference to the taste of the bread.

Adding fat—oil, butter, margarine, shortening, lard, etc.—will prevent the gluten from forming large sheets. The fat gets in the way of the small sheets joining up; it "shortens" the strands and sheets of gluten, hence the word *shortening*. Adding shortening makes the result more cakelike and less breadlike. Some recipes have you oil the outside of the dough to keep it from sticking to pans, fingers, and breadboards. Others have you paint melted butter on top of a baked loaf to keep the crust from getting dry and hard. Neither of these uses has much or any effect on the interior of the loaf.

Bread flour is flour grown and processed to contain a lot of gluten. Cake flour is designed to have less gluten. All-purpose flour is a mix that has an intermediate amount of gluten. If you are using all-purpose flour, you probably won't have much use for shortening in bread dough, but shortening will make your cakes more tender and cakelike, though less breadlike.

The chart on the next page shows 10 simple bread recipes collected from different sources, with the ingredient amounts converted into percentages to make it easy to see how variable a basic bread recipe can be and still end up producing very similar results. The last column in the chart is the average recipe.

Converting the average recipe into a one-loaf batch by assuming 2 teaspoons is 1 percent and rounding the numbers, you have:

- 144 teaspoons flour (3 cups)
- 56 teaspoons water (1⅛ cups)
- 1 teaspoon sugar

10 Simple Bread Recipes

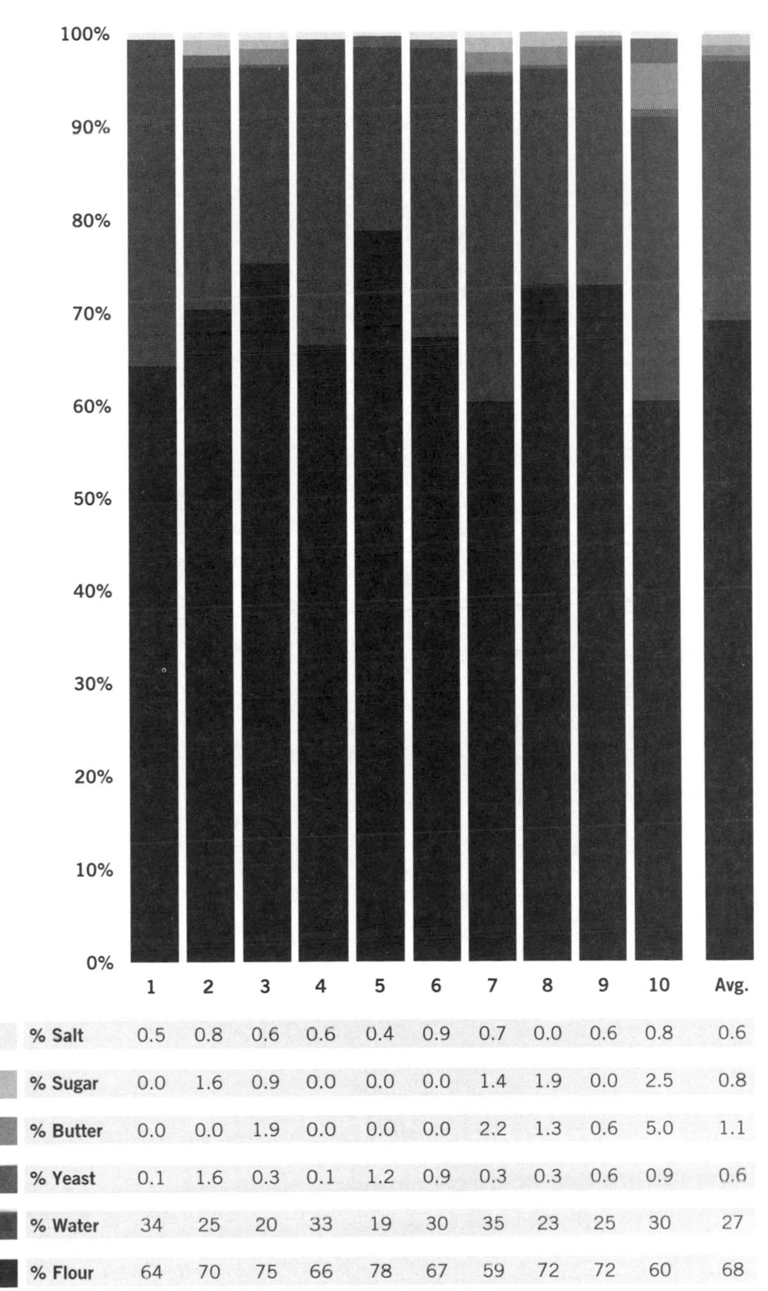

	1	2	3	4	5	6	7	8	9	10	Avg.
% Salt	0.5	0.8	0.6	0.6	0.4	0.9	0.7	0.0	0.6	0.8	0.6
% Sugar	0.0	1.6	0.9	0.0	0.0	0.0	1.4	1.9	0.0	2.5	0.8
% Butter	0.0	0.0	1.9	0.0	0.0	0.0	2.2	1.3	0.6	5.0	1.1
% Yeast	0.1	1.6	0.3	0.1	1.2	0.9	0.3	0.3	0.6	0.9	0.6
% Water	34	25	20	33	19	30	35	23	25	30	27
% Flour	64	70	75	66	78	67	59	72	72	60	68

- ½ teaspoon salt
- ⅔ teaspoon yeast
- 1 teaspoon melted butter

Saving the butter for painting on the top of the loaf when the bread has baked, mix all the other ingredients in a bowl until they are well blended.

The next step is to help to process the gluten. There are some recipes that do not call for kneading the bread. These recipes let time do the work for you. You simply let the dough work all by itself overnight and most of the next day, all alone in its bowl, covered by a damp towel.

Most recipes, however, assume you want your bread the same day. To speed up the formation of the sheets of rubbery gluten, set the dough onto a floured board (so it doesn't stick to the board). Fold it over and press it flat repeatedly, for something like 8 to 10 minutes, adding flour as needed to prevent it from sticking.

This gluten-forming process has the unfortunate side effect of removing many of the bubbles that may already have been in the dough. To get back the bubbles you need, let the bread sit in a place where it won't dry out, usually a greased bowl covered with a damp towel.

Let the dough rise for 15 minutes, or until it has doubled in size. The actual length of time here is not very important. Recipes vary quite a bit. Some have you let the dough double, then punch it down and let it double again to form more gluten. How much you play with the dough will depend on how much gluten the flour has and how much gluten you want to develop.

But the result, whether it is a soft, light, cakelike loaf or a rugged, firm, hearty loaf, will still be recognized as bread.

Now it's time to bake the loaf. It can be placed in a loaf pan or just on a greased cookie sheet. You can decorate the top by scoring it with a knife or form the dough into ropes and braid them together. All of these are just decorative variations.

The oven for bread is generally hot, about 400°F (200°C) or even hotter. The time it takes to bake is 40 to 50 minutes, or until you like the color of the crust.

When the baking is done, brush the top with the melted butter. If you like a soft crust, you can let the loaf cool in a plastic bag. For a dry, hard crust, just let it cool on the countertop.

Some recipes call for placing a pan of water in the oven along with the loaf, to keep the crust thicker and crisper. This also speeds the baking (moist air conducts heat faster than dry air). This is optional. The steam condenses on the cold dough at first, which slows the formation of the crust. Some of the sugars in the dough dissolve in this condensed water layer, which contributes to browning. A lot of steam will make the crust thicker, shinier, and a darker brown.

Now that you know why the ingredients are there, and why the processing steps are needed, you can throw away your measuring equipment and do the whole thing by eye.

Dump some flour into a bowl. Add some yeast. Add some water gradually, stirring until the dough is about the consistency you remember from the times you made bread from a recipe. Knead the dough for a while on the floured cutting board. Let it double in a greased bowl with a damp cloth cover. Form a loaf on a baking sheet, preheat the oven to 400°F (204°C), and bake until you like the color.

That's how bread was made for centuries before the invention of measuring cups and ovens that kept a steady temperature. Usually the yeast was just a bit of bread dough saved from the last batch. Add a little water and sugar, and your "starter" will grow plenty of yeast for the next bread-baking session.

Leavening Alternatives

Yeast is a convenient leavening agent (something that makes bubbles of gas in a dough). Yeast spores float in the air and form white films on grapes and plums and other fruits with thin skins that allow sugars to get to the surface. You can make yeast starter for your own wine, beer, or bread by culturing the white film from grapes in a little sugar water.

Some breads, however, just use steam and hot air for leavening. Popovers are an example of steam-leavened bread. But the prize for steam leavening goes to popcorn. We go to all the trouble to grind wheat, add yeast, knead, and bake just to get a foam made from seeds, while popcorn does it fine with just some heat. Puffed wheat and puffed rice are made by heating the seeds under steam pressure (in a big pressure cooker called a "gun") and then suddenly releasing the pressure (called "firing the gun"). This whole process takes less than a minute.

So-called quick breads (because you don't have to wait for the yeast to grow or the gluten to develop) use baking soda and an acid, such as buttermilk, to form bubbles of carbon dioxide gas. Since these breads are not kneaded or left to themselves overnight, they have little gluten and are more like cake than a sturdy loaf.

Baking soda is sodium bicarbonate:

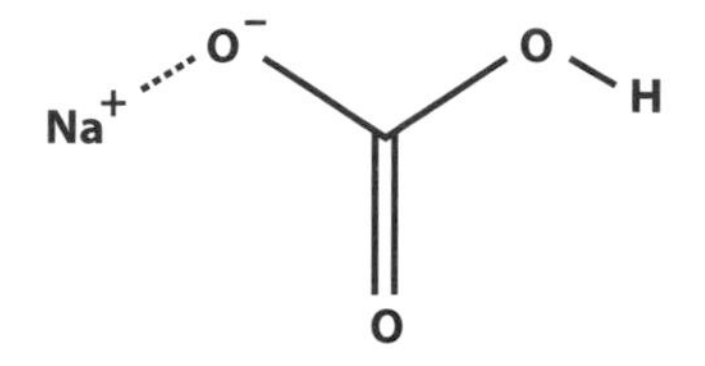

There is a carbon atom in the middle, with three oxygen atoms, a hydrogen atom, and a sodium atom. When sodium bicarbonate is added to water, it breaks apart into three ions. These are a positively charged sodium ion, Na^+, a negatively charged hydroxide ion, OH^-, and carbonic acid, which is what we call soda water (carbon dioxide dissolved in water).

If you add an acid, such as vinegar (acetic acid), a reaction occurs and you get sodium acetate and water as products when the acid reacts with the sodium and hydroxide ions. What is left is carbonic acid. Carbonic acid (carbonated water) fizzes, releasing bubbles, just like in soda.

Any common acid will react with baking soda this way, so the lactic acid in buttermilk, the citric acid in lemon juice, or the acetic acid in vinegar can be used, too.

But bubble formation is only half of what happens in the bread. As you bake the bread, the heat causes the gas to expand. The heat also denatures the proteins and starches, allowing them to link up into solid webs, holding the shape of the bubble even after the gas has cooled. Air from the room slowly fills the spaces where the steam and hot expanded gas used to be, as the bread slowly cools.

If the baking has not completed, and the proteins are not yet firm, then allowing the temperature to drop can cause the

Chemistry Lesson

Ionic Bonds

In this book we will be dealing with three types of chemical bonds—ionic, covalent, and hydrogen bonds.

An ion is an atom that has a charge, because it has either gained or lost one or more electrons.

Ionic bonds form when a metal like sodium loses an electron to an atom of something such as oxygen or chlorine, because oxygen and chlorine have a higher affinity for electrons than sodium does. This makes the oxygen or chlorine atom end up with a negative charge, leaving the sodium atom with a positive charge.

Since opposite charges attract, the sodium atom hangs around near the atom that took its electron, and this attraction is called an *ionic* bond.

bubbles to all shrink back down, and we say that your soufflé has "fallen." This can happen with breads and cakes as well as fluffy egg dishes.

But firm protein networks are not the whole story. During baking, what started out as a foam (a collection of closed bubbles) becomes a sponge (where all the bubbles have broken to form an open network that air and water can flow through). The heat has not just firmed the proteins and made them bond together, but it has also expanded the gas in the bubbles to the point that the bubble walls have broken, letting the gas escape. This is important, because otherwise the gas would cool and contract, and the resulting vacuum would crush the bubbles back into a dense mass of dough.

To understand this better, put a little bit of water into an empty aluminum soda can and heat it on the stove until it is full of steam and no more water is left inside. Using a hot pad or tongs so you don't get burned, quickly invert the can into a saucer of water so that the opening is underwater. The steam will cool and contract, which draws the water into the can. This cools the can so quickly that the water can't get into the can fast enough to prevent the weight of the miles of air above it from crushing the aluminum can.

If a bubble of aluminum can't withstand the weight of the air, neither can a simple bubble made of starch and gluten. However, the open network of a sponge, where all the bubbles have been broken, lets the air in, and nothing gets crushed.

Gelatin Foam

Other proteins can also make foams. One simple protein is gelatin, which is used to make marshmallows.

Marshmallows are made by cooking a sugar syrup to the firm ball stage (240°F, 116°C) and then beating that into gelatin that has softened in cold water.

As with other protein foams, the gelatin will denature as the hot syrup and whisking cause it to form links with itself, forming a sturdy net. The syrup attracts the water-loving parts of the protein, leaving the oil-loving parts facing the air in the bubbles.

You can divide a marshmallow recipe into steps, then look at why each step is there. The first step is to sprinkle gelatin powder into cold water. This is done to soften the gelatin without changing the shape of the protein molecules. You only want them changing shape when you beat in the syrup.

The next step in the recipe is to make the syrup. Most candy recipes are primarily about controlling the crystallization of sugar. This is done by controlling the concentration of sugar in the water, making sure there are no seed crystals falling into the syrup to prematurely begin crystallization, and providing simple sugars in addition to the sucrose (a complex sugar made up of the two simple sugars glucose and fructose).

Simple sugars bind more tightly to water than sucrose does, and so they don't crystallize as easily. There are two basic ways to get simple sugars into a syrup. The first is to simply add them. A candy recipe that calls for corn syrup in addition to sugar is doing just that; corn syrup is mostly simple sugars.

The second way is to break up the sucrose into its two simple sugars. You do this by heating it in the presence of an acid. A candy recipe that calls for cream of tartar (tartaric acid) is doing that. Other recipes might call for vinegar, lemon juice, or other acids to cook with the sugar.

The third step in a marshmallow recipe is when the foam is actually made. Generally, syrup is slowly added to the gelatin mixture while the beaters are running at medium speed (to

Chemistry Lesson

Covalent Bonds

Electrons in an atom can't all share the same space. Close to the nucleus, there is room for two electrons. If those two places are filled, a third electron will have to start filling the next shell out from the nucleus. That second shell is bigger and can hold eight electrons.

A hydrogen atom has only one electron, so there is room for another electron in its innermost shell. When a second hydrogen atom gets very close to the first, an electron from each atom can fall into the space in the other atom's shell, so that each atom has two electrons filling its innermost shell.

Like a domino that has fallen over, electrons that have fallen into an inner shell will stay there unless energy is added to pull them back out. The hydrogen atoms have joined (bonded) to make a hydrogen molecule, and they will stay stuck together until enough energy has been added to pull them apart.

When electrons are shared between two atoms in this way, it is called a covalent bond. Covalent bonds can occur anywhere there is an unfilled shell that electrons can fall into. Carbon has four empty spaces in its outer shell, so it can form four covalent bonds with other atoms. When it does this with four hydrogen atoms, the molecule is called methane, or CH_4.

prevent splashing), and then the speed is turned up to high for a good 15 minutes or so, to denature the proteins and form a stable foam. The foam is then turned out into a greased pan to sit for half a day—8 to 12 hours—to let the protein net complete bonding.

Often the bulk of the recipe deals with controlling the stickiness of the whole mess. Pans are greased, lined, or lined and greased. A mixture of cornstarch and powdered sugar is spread out over a cutting board and the cooled foam is dumped out of the greased pan onto it, and the powder is generously applied to the cut surfaces after the candy is divided into bite-sized pieces. The knife used for the cutting is coated with vegetable oil for the first cut, and then coated with powdered sugar and cornstarch for subsequent cuts.

Marshmallows were originally a way of making the cough suppressant in the root of the marshmallow plant palatable. The starch in the root was mixed with egg white as the protein, and the syrup was beaten in. Some recipes still mix in beaten egg whites with the gelatin foam.

Sugar Foam

You can make foam without protein. The suds in dishwater are made of just water and detergent. The molecule in detergent, like protein, has a part that loves water and a part that avoids water in favor of air or fats. It is used to clean up greasy dishes because the fat-loving ends attach to the fat, while the water-loving ends prevent fat globules from coalescing together by making a protective coat around them. But soapsuds are not a stable foam. Eventually, all the bubbles pop, and you are left with plain dishwater again.

But if you make a foam out of something that hardens when it cools, then you don't need a protein to make the foam stable. A kind of candy called honeycomb is just that sort of foam. You start by cooking a sugar syrup (sucrose and simple sugars) to 300°F (150°C), the so-called *hard-crack stage*. This is the stage just before the sugar burns (caramelizes) and turns brown. When this syrup cools, the result is hard candy, like a lollipop.

As in any candy cooking, you must be careful not to let any sugar crystals fall into the syrup from the side of the pan. This will cause premature crystallization, resulting in a grainy texture instead of a glassy one.

When the syrup has reached 300°F, you add baking soda and quickly beat the syrup. The high heat causes the baking soda to decompose into sodium carbonate and carbon dioxide bubbles. Note that this happens with heat alone, and unlike earlier, in the discussion about quick bread (page 21), no acid is needed. The syrup will foam up and triple in volume as you whisk it.

The last step is to pour the foam out into a pan prepared (as with the marshmallows) with grease, lining, or both, to prevent sticking.

This kind of candy will absorb moisture from the air easily and become a soggy mess, unless it is coated in chocolate or kept in a sealed container. Of course, you could always just eat it right away.

Recipe

Whipped Creamsicle Topping

Did you love Creamsicles when you were a kid? Vanilla ice cream and orange popsicle together in the same bar, like an orange cream soda on a stick. With ice cream!

Whipped cream is great fun all by itself. But sometimes you just want to surprise your dinner guests with something out of the ordinary. Whipped cream that tastes like an orange Creamsicle goes great with pumpkin pie or ice cream.

You can do a couple of interesting things to make whipped cream a little different from what you might be used to. Add a tiny amount of xanthan gum to make the foam stiff and long lasting. And make it in a neat whipping gadget so you can make it up days before you need it, and keep it in the fridge until you need it. Who wants to be whipping cream when you could be talking with your guests after dinner?

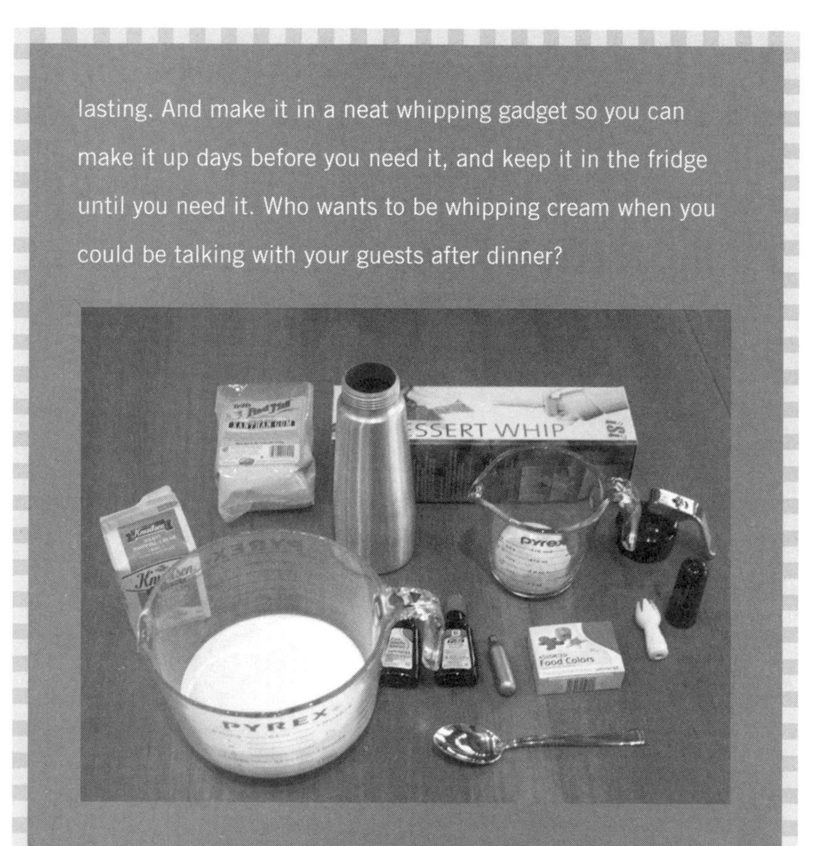

The dessert whipper shown was purchased at Amazon.com. There are one-cup sizes and one-pint sizes. This recipe uses the one-pint size.

Ingredients:

- 1 pint heavy whipping cream
- 2 teaspoons vanilla extract
- 2 teaspoons orange extract
- Food coloring (red and yellow) to make a nice orange color

- ¼ teaspoon xanthan gum
- ¼ cup sugar

Supplies:

- 1-quart bowl
- Spoon
- Dessert whipper
- Whippits

Start by pouring the whipping cream into a quart bowl. Add the vanilla and orange flavoring first, since they will change the color of the cream a bit.

Add the food coloring after the flavorings to get the color just the way you want it. I like twice as much yellow as red,

16 drops of yellow and 8 drops of red, but make it whatever color looks best to you.

Next, add the xanthan gum to the *sugar*, not to the cream. Stir it well into the sugar to ensure that it will dissolve well in the cream and not clump up.

When the xanthan gum is mixed well with the sugar, stir the sugar into the cream. Stir the cream well, until all the sugar has dissolved. When there are no longer any grainy bits of sugar in the cream, pour it into the dessert whipper.

The whipper is charged with small cylinders of nitrous oxide gas called *whippits*. Buy them when you buy the dessert whipper, since the two are sold separately. One whippit is good for one cup, so you will need two for the larger dessert whipper. Charge the whipper with one cylinder, and shake 10 to 12 times. Then remove the whippit and replace it with a fresh one.

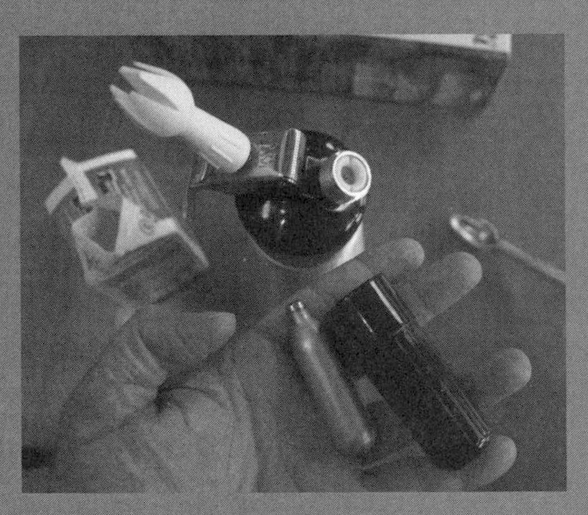

Let the nitrous oxide dissolve in the cream for a few minutes in the refrigerator before trying to dispense it. You can keep it in the dessert whipper in the refrigerator for a week past the "sell-by" date on the carton of cream.

A Note About Xanthan Gum

While creating this recipe, I tried several ratios for xanthan gum to cream. The recipe works well for amounts as little as 1/8 of a teaspoon to about 3/8. When I used a full teaspoon of xanthan gum, the mixture was very thick and hard to stir, and it whipped up into the most delicious foam rubber I had ever tasted. Most of the mix remained inside the dessert whipper and had to be spooned out.

Whole Foods sells xanthan gum, and many gourmet cooking stores, such as Sur La Table, sell the dessert whippers.

Of course you can just whip the cream using a whisk or a kitchen mixer, but then you wouldn't have an excuse to buy the fun toy!

A Note About Nitrous Oxide

Whippits are filled with nitrous oxide (N_2O). This is also known as laughing gas, the anesthetic used by dentists to put you to sleep. It is used in whipping cream because it dissolves easily in butterfat, it does not allow the fat to oxidize and get rancid as compressed air would, and it does not curdle the milk proteins as carbon dioxide would.

❖ 3 ❖

EMULSIONS

Oil and water may not mix, but with a little help, the two can join together in an *emulsion*. You can have an emulsion of fats in water, such as milk, cream, or most salad dressings, or you can have an emulsion of water in fat, as you do in butter and peanut butter.

Why Some Things Don't Mix

It is common to say that oil and water don't mix, but lots of things don't mix. Air and water, air and oil, sand and water, sand and oil, sand and air—all of these will separate into layers if you put them in the same container.

What makes things separate is the attraction that molecules have to each other. Water molecules bind to one another using bonds that are stronger than the ones that air molecules bind with.

So the water molecules join up and leave the air behind. Water molecules by themselves are lighter than air (a water molecule has a molecular weight of 18, while air is made of nitrogen molecules that weigh 28, and oxygen molecules that weigh 32). But when the water molecules bind together, the bond is strong enough to pull the molecules close, and the density goes up, and it rains.

The molecules in sand or steel bind even tighter than water, and these materials become dense solids. The molecules in oils and fats bind to one another only with weak bonds. As with air, the water molecules bind together, leaving the oil behind. Oil and fat molecules have long chains of atoms that are bulky and tangle together, making them less dense than water, more viscous, and in the case of fats with really long chains, solid.

Emulsifying Agents

When you make foams, you get air and water to mix and stay stable. You can use the same tricks to make oil and water mix and stay stable—just use a molecule that has both a part that likes water and a part that doesn't.

Proteins work, and so do smaller molecules, such as soap and detergent. A soap is basically a fat attached to a water-loving element, such as sodium or potassium. While good at making emulsions in the dishwater, soaps have a flavor most people dislike and are not used in fancy French sauces.

Luckily for the cook, plants and animals find these double-ended molecules very handy as well, and they produce them in large quantities. Cell walls are made of things called lipid bilayer membranes. These are sheets of molecules called *phospholipids*.

Chemistry Lesson

Hydrogen Bonds

Some atoms, such as chlorine, fluorine, oxygen, and nitrogen, have strong affinities for electrons. When these atoms covalently bond to hydrogen, the electrons seem to spend more time near the heavier atom than near the hydrogen.

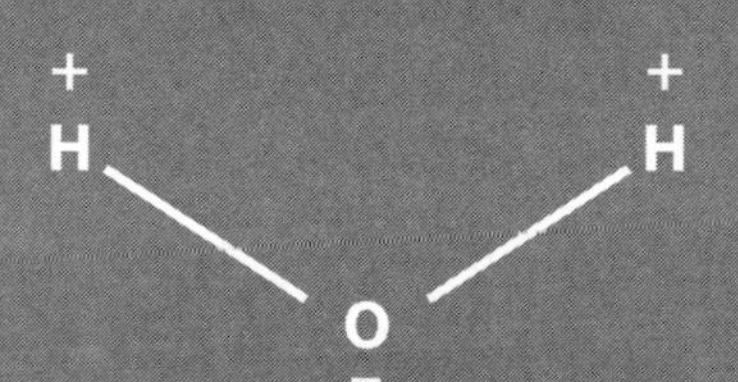

This makes one part of the resulting molecule a little bit positive, and the rest a little bit negative. When molecules like this get together, their negative sides are attracted to one another's positive sides, and a weak bond is formed. This is called a *hydrogen bond*.

Water forms hydrogen bonds because the oxygen side is more negative than the side that has the hydrogens. The water molecules can also form hydrogen bonds with other molecules that have similar areas of different charges. Molecules that have a positive side and a negative side are called polar molecules.

They have water-loving sides and water-avoiding sides. In water (in the cell) they join up back-to-back, keeping their water-loving sides facing the water—thus the "bilayer" part of the name.

If you grind up the cell walls, and mix in a little bit of oil, you can get those layers to open up to make a single membrane around the oil droplet, with the water-loving side out, facing the water. This water-loving side prevents other oil droplets from combining with this one.

A common phospholipid is lecithin. It is found in the yolk of eggs, but commercially it is extracted from soybeans, which are cheaper. But there are phospholipids in almost every living cell, whether plant or animal.

When you mention emulsion to a cook, he or she will think of mayonnaise. In mayonnaise, the emulsion is stabilized by the lecithin and proteins in the egg yolk, and the phospholipids in the ground mustard. Other emulsified sauces are stabilized by the phospholipids from garlic or some other plant material.

Gum Stabilizers

Another stabilizer for emulsions is vegetable gum. Gums are starchlike large molecules that form thick colloidal suspensions in water. These have the effect of keeping the oil droplets from recombining simply by getting in the way, forming semirigid walls between the droplets. The gums in mustard and in garlic help to stabilize emulsions that contain them.

Sometimes it seems like people want to make traditional emulsion sauces just because they are notoriously hard to make. Using a double boiler to prevent the eggs in a hollandaise sauce from scrambling is considered cheating by some, despite the results being identical to the traditional method. Adding a pinch of xanthan gum when making a beurre blanc sauce will prevent it from separating, but purists would never hear of doing so.

The advantage to the practical cook is that you can use whatever methods you like, and then name the resulting sauce after yourself, even if it tastes just like hollandaise or béarnaise sauce.

Shortcuts and Aids

There are plenty of recipes and long discussions about how to make traditional sauces in the traditional ways. I'd rather talk about the cheats.

Getting the emulsion started is one common problem. You are advised to add the oil to the beaten eggs and mustard very slowly at first, and to add only a small amount of the acidic vinegar or lemon juice at first as well. Later, when the emulsion has started, you can pretty much dump oil and vinegar in wholesale. But I have never heard anyone suggest the obvious solution: just start with a dollop of yesterday's mayonnaise beaten into the egg and mustard, then dump in the other ingredients. That works.

Adding ⅛ of a teaspoon of xanthan gum to any oil and water emulsion will remove most of the barriers to success. You can be much more relaxed about amounts and rates, and it will be much more resistant to separation.

Some emulsions, such as hollandaise sauce, call for cooking an egg-stabilized emulsion. The proteins in an egg are easy to denature with heat. That is what you see happening when an egg white turns from transparent to actually being white. The egg proteins start out carefully folded up so that they can do the job they do in the cell. You heat them to get them to open up, so that they tangle together and prevent the flow of water in the mixture, and so the oil-loving parts can find the oil. If you heat them too much, they will start bonding together into a strong rubbery network, giving you scrambled eggs instead of a thick sauce.

Eggs begin to coagulate at 160°F to 170°F (70°C to 77°C). Not coincidentally, this is also the temperature needed to coagulate the proteins in salmonella and other pathogens, thus killing them. In order to cook the eggs well enough to kill bacteria, but still prevent them from scrambling, you can use a well-known trick from both chemistry and your mother's cookbooks: add an acid to the eggs.

Acids prevent some of the chemical bonds from forming between the proteins until the temperature gets much higher, closer to 195°F (90°C). So adding some lemon juice or vinegar to the sauce will prevent the eggs from curdling, at least if you keep the temperature well below boiling.

Hollandaise Sauce

Let's walk through the creation of a hollandaise sauce and see what steps there are and why they are needed.

Some recipes start with *clarified butter*, while others use whole butter. Clarified butter is butter that has been heated until the emulsion breaks and the water and milk solids fall to the bottom, leaving pure butterfat at the top.

If the recipe uses clarified butter, you lose the parts of the butter that help it emulsify, and you lose the flavor elements of the milk solids, as well as some important water. An emulsion needs water as much as it needs oil. So a recipe using clarified butter will need more water or lemon juice than one that does not.

To clarify butter, the heat should be very low. You just want to separate the emulsion; you don't want to burn the milk solids or boil the water in the butter. There will be some froth on the top of the clarified butter. This is discarded, along with the milk solids and water, since you only want the butterfat in clarified butter.

While the butter is clarifying, you make a reduction of white wine vinegar, crushed white peppercorns, white wine, and minced scallions. Simmer these until you have reduced the liquid by half, to about a tablespoon or two.

Once the butter is clarified and the wine reduced, bring water in the bottom pot of a double boiler to a simmer. The water should not be high enough to touch the top pot of the double boiler. You only want steam touching the top pot. This will keep the temperature low enough to prevent the eggs from curdling.

When the water has come to a simmer, take the pot off the burner. Put two egg yolks into the top of the double boiler, along with a tablespoon of the reduction. Immediately start whipping the egg yolks. Whip the yolks until they lighten to the color of butter, and they start to thicken.

Now you start drizzling in the butterfat. Just a few drops at first, allowing the butter to be absorbed as you continue to whip the eggs. When the eggs have absorbed about 4 ounces (8 tablespoons) of butterfat, you have started the emulsion, and you can add the rest of the butterfat more quickly.

Once all of the butterfat has been whisked in, you can add other flavorings, such as lemon juice, cayenne pepper, Tabasco sauce, or Worcestershire sauce, to your taste.

The sauce is generally served right away, since it must be kept warm enough to prevent the butterfat from congealing and separating the emulsion. This obviously means that it can't be refrigerated, so there is no safe way to keep it for another day.

Other Emulsifiers

Lecithin, the phospholipid found in eggs and cell walls, is one example of an emulsifying agent. We have also discussed proteins and detergents. Another class of emulsifier comes from taking apart a fat molecule (or not completing the building of a fat molecule).

Fats are *triglycerides*. This means that they have glycerin as a backbone, and attached to the glycerin are three (hence the "tri-") fatty acids. If you remove one or two of those fatty acids, then part of the glycerin molecule is left available for attaching to something else, such as water. This would leave a molecule that has a water-loving end (where the glycerin is) and a fat-loving end (the remaining tails of the fatty acids). This would make a good emulsifying agent.

These partial fats are called *monoglycerides* or *diglycerides*, depending on whether there are one or two fatty acids attached to the glycerin. You may have read a food package label that mentions mono- and diglycerides. Now you know that they are there to stabilize emulsions or foams. They are just partial fats. Glyceryl stearate is an example you might see on a package label.

Various polysorbates are also commonly used as emulsifying agents. Again, a fatty acid is attached to a water-loving molecule. Polysorbates come in a wide variety, and the number after the name indicates the length of the fatty acid chain. Polysorbate 20 and polysorbate 80 are common examples; the latter is used in ice cream to modify how the proteins coat the fat droplets. Polysorbate 60 is used in hot cocoa mix.

Similar compounds are ceteareth alcohol, cetyl alcohol, or stearyl alcohol. These are compounds known as emulsifying waxes.

Emulsifying agents generally dissolve better in one part of the emulsion than the other. Molecules such as proteins, which dissolve better in water than in oil, will help to make oil-in-water emulsions, such as milk and cream. Molecules that dissolve better in fats and oils help to make water-in-oil emulsions, such as butter and margarine. Churning cream helps to denature some of the proteins in the cream until they fold in a way that makes them dissolve better in fat, and the emulsion becomes one of water in solid butterfat, and you get a solid.

To make a water-in-oil emulsion like margarine, you would use an emulsifying agent that dissolves better in fat. Something with a longer fatty acid chain would dissolve better in fat than one with a shorter fatty acid chain.

❖ 4 ❖

COLLOIDS, GELS, AND SUSPENSIONS

Emulsions are just one example of a *colloid*, a mixture in which one substance is evenly dispersed in another. The particle size of the substance that is dispersed is too large for it to dissolve and too small for it to settle out easily. Generally this means the particles are between 5 and 200 nanometers in diameter.

Fog and clouds are colloids of water in air. Smoke is a colloid of solid particles in air. As you have read, foams are colloids of a gas in water, and emulsions are colloids of one liquid in another. Ink is a colloid of a solid pigment in water. Styrofoam is a colloid of a gas in a solid. Gels are a colloid of a liquid in a solid. There are also colloids of solids in other solids, as in some types of glass.

Water-Based Colloids

One class of colloid often encountered in cooking are the *hydrocolloids.* These are gels (solid) or sols (liquid) made of particles dispersed in water. Gelatin is a sol when hot and a gel when cooled. Other examples are jellies made from pectin, agar, carrageenan, or other gelling agents.

If the particles are too large, instead of a stable colloid it's a *suspension.* That's when particles are temporarily suspended in a fluid but will eventually settle out if left undisturbed. The particle size is generally larger than 1 micrometer. Particles between 0.2 micrometers and 1 micrometer can form colloids or suspensions, or something that behaves somewhat like either one.

In the kitchen, cooks often go to some effort to create colloids, and at other times to prevent them. They add pectin to fruit juice to make jelly, but add flocculating agents to wine to clarify it by removing fine particles. Particles in water often form colloids when they all carry the same electric charge and thus repel one another. Adding salts to a colloid can provide charged ions that surround charged solid particles and neutralize the repulsion. This allows them to bind together into particles too large to remain as colloids, and they settle out.

A gel is formed when chemical bonds cause the particles to cross-link into a three-dimensional network that behaves like a solid, even though it is mostly liquid. Cooked egg white is a gel formed when the proteins in the egg white open from their natural folded shape and then bind to one another in many places.

Proteins are large molecules that, when opened up and spread out, provide lots of opportunities to bind and form links to one another. Other large molecules that form gels include starches and charged polymers.

Starches

Starch is a polysaccharide, a word that simply means "many sugars." It is formed when molecules of the simple sugar glucose are joined together to form long chains and branched trees. The long chain form is called amylose, and the branched form is called amylopectin. In most plants, starch grains are about one-quarter amylose and three-quarters amylopectin. Animals form a more branched type of amylopectin, called glycogen, to store energy.

In the organism creating the starch, there are enzymes that create straight chains (amylose), enzymes that create branched trees (amylopectin), and enzymes that convert branched tree shapes into long chains. All of these starch forms are used for different purposes in the cell. Some are structural and some are related to energy storage.

In a plant, starch is bundled into crystallized packets called starch granules. The long chains align and pack tightly, creating a dense solid. When heated in water, these starch granules absorb water and expand, eventually separating into a loose liquid colloid. If left to cool, the chains in the starch once more align and form a more solid gel, at the same time excluding water in a process known as syneresis.

Amylose molecules are smaller than amylopectin molecules. An amylose molecule might be composed of 250 to 2,000 glucose molecules and have a molecular weight of 40,000 to 340,000. Amylopectin is composed of many branching chains of amylose, and it can have a molecular weight of as much as 80,000,000.

Amylose often forms double helices that form long hollow tubes. In one common test for starch, iodine is added to a starch solution. The iodine fits neatly inside the helical tube, changing the way it interacts with light. The result is a blue-black color in a solution that was once clear.

Agar and Agarose

Agar is a gelling agent that is similar to starch. Instead of using glucose molecules as a base, the simple sugar galactose is used. As with starch, there is a long-chain version, called agarose, and a branched version, called agaropectin. Of the two, agarose is the primary gelling agent; purified agarose is widely used in molecular biology because it has excellent gelling characteristics, and it has large pore sizes, which allow easier transport of molecules through the gel.

Agar has advantages over starch in both cooking and science. It melts at a high temperature (185°F, 85°C), but solidifies at a much lower temperature (90°F, 32°C). Water molecules bind inside the helices of agarose and stabilize the gel, creating a strong gel without losing any water. Bacteria can't eat agarose like they can starch, so it is used to culture colonies of bacteria on agar plates (which can be incubated at higher temperatures than starch can, due to the higher melting point).

In cooking, agar is used as a vegetarian gelling agent instead of gelatin. Agar is extracted either from red algae or from a type of seaweed. Similarly, another gelling agent, carrageenan, is extracted from certain red seaweeds. Like agar, carrageenan is formed of galactose and curls into helices, making it gel nicely at room temperature. It is viscoelastic like toothpaste, meaning it can liquefy under shear stress and be pumped or extruded easily, and then regains its solid form when the stress is removed.

Pure agarose makes excellent gels. It is used in molecular biology to separate strands of DNA by their sizes. The large, shaggy agaropectin molecules would get in the way, but the pure agarose allows the DNA to flow through the gel under the influence of an electric field.

Pectin Gels

Pectins are one of the soluble dietary fibers, and as such they have well-known effects on human digestion as well as lesser-known effects, such as reducing the digestibility of some proteins and amino acids. The remedy Kaopectate used to be made of pectins and a fine clay called kaolin.

The big, shaggy pectin molecules are used most in making jams and jellies. Pectins used for making jelly come from citrus fruits and apples, but most plant cell walls contain pectins as structural elements and for lubrication. Pectins hold cell walls together. When fruit ripens, enzymes are released that break down pectin and make the fruit get soft. The pectin in jams and jellies is for the most part indigestible, although some gut bacteria can metabolize it.

Generally, harder fruits such as apples, oranges (the peel), and plums have the most pectin, while softer fruits such as grapes, cherries, and strawberries have much less, since pectin is a structural molecule that binds the cells together. Citrus peels are about 30 percent pectin, and whole oranges are about 2 percent. While whole apples have even less (1 percent), commercial pectin is extracted from the concentrated solids left over after pressing the juice from apples.

Commercial pectin comes in two forms: high-methoxyl pectin (HMP) and low-methoxyl pectin (LMP), and they set into gels in different ways. Traditional pectin is the high-methoxyl type, which gels only when the sugar content is very high (generally more than half sugar by weight) and the acidity is high.

HMP molecules are negatively charged and repel one another. This means they won't develop bonds between the molecules and form a solid gel. Two things are needed to overcome this. One is acid, which neutralizes the negative charges. The other is a high concentration of sugar, which binds well to water molecules, so that the water binds less to the pectin. This allows the pectin molecules more opportunities to bind together.

LMPs were developed to allow gelling in the absence of sugar. They are usually made by treating HMPs with ammonia, sodium hydroxide, or acids. Instead of requiring sugar and acid, low-methoxyl pectins require a source of calcium to gel. Generally, a slow-dissolving calcium salt is used, such as calcium phosphate, to give the pectins time to fully dissolve before they react with the calcium.

Protein Gels

You are already familiar with several protein gels. Scrambled eggs and yogurt are gels formed by proteins. The words *gel* and *gelatin* both come from the Latin word for "freeze," and gelatin is a protein gel.

Water-soluble proteins such as the albumins in egg whites are generally spherical globs, with the hydrophobic (water-avoiding) amino acids tucked away inside. They can form gels and colloids in this state (as the ovomucin does in the firm gelled part of the egg white). But when they are denatured by heat or beating or acids, they unfold and can join together to form firmer gels, or tough plastic sheets.

Gels made from proteins are very common in the kitchen. From scrambled eggs to yogurt, custard to gelatin, proteins form gels when heated or acted upon by acids or enzymes.

When proteins are heated, they open up from their natural folded state and take up more room, impeding the flow of the water they are in, causing it to thicken. With further heating, the proteins bond together, forming a firm solid gel.

The nature of the protein determines features of the gel. Egg protein has many sulfur-containing amino acids, and the sulfur atoms find it easy to link together to form disulfide bonds. This cross-linking firms up the gel, which is the reason why egg proteins are so easy to denature into a gel that they can be cooked on a hot sidewalk in the summer.

Milk proteins are easily denatured by acids. As the bacterium *Lactobacillus acidophilus* oxidizes the milk sugar lactose into lactic acid, that acid unfolds the protein called casein in the milk. The proteins can then slowly link together to form the soft gel we call yogurt.

Gelatin is a protein extracted from the collagen tissue that connects bones to each other and to muscle. It makes up most of the protein in skin and bones.

Collagen is insoluble, made up of three chains of protein that twist into a rope in what is called a triple helix. When heated in water, usually with added acid or alkali, it hydrolyzes, and the helical rope unwinds, letting the individual chains float free.

When the solution cools, the chains begin to curl back up into helices. However, they are all tangled together, and as they curl, the tangling increases, like a coiled telephone cord. The chains knot together, and some of the ends of the chains form double and triple helices with one another.

All of this tangling forms a kind of gel that is different from what you have read about in other protein gels. Gelatin forms a thermo-reversible gel. Since the chains don't form a lot of strong cross-linked bonds, the gel can be undone by reheating. This makes it similar to gels made from vegetable gums and starches.

Gelatin melts at body temperature, and this gives it special qualities in foods that melt in your mouth.

Recipe

Cherry Dream Cheese

Making your own cheese is fun, but since a gallon of milk costs almost as much as a pound of cheap cheese, and almost half as much as a pound of good cheese, the cheese you make has to be pretty good to justify the expense.

This cheese fits the bill perfectly.

Cheese excites the salt sensors on the tongue, as well as the savory, or umami, sensors that respond to glutamate, indicating a rich protein source.

To round out the flavor palate, it's good to excite the sweet sensors and the sour sensors on the tongue as well. (I'll let someone else devise a good-tasting recipe that also excites the two bitter sensors—I'd just as soon avoid those myself.)

Dried Bing cherries are just the ticket. Not so sweet as to overpower the cheese, not so tart as to pucker faces, these colorful little nuggets nestle into the cheese as if they belonged there. And they do. This recipe is for about a pound of finished cheese.

Ingredients:

- 1 gallon whole milk
- ~½ cup unflavored (or vanilla-flavored) yogurt or cultured buttermilk
- ~¼ tablet rennet
- 2 teaspoons salt
- ½ cup dried Bing cherries
- 1 tablespoon sugar
- 1 red crayon
- ¼ pound paraffin wax

Supplies:

- 6-quart pot with lid
- Long-handled spoon
- Long knife
- Large strainer

- 1 foot of 4-inch-diameter ABS pipe, cleaned well
- 4 square feet (or more, if you like) of cheesecloth (I prefer the fine-weave type)
- Disk of wood or hard plastic to just fit inside pipe (I cut the plastic lid from a large container of chocolate-coated almonds)
- 50 pounds of something (concrete blocks, bricks, old gym weights, etc.)

Start out by sterilizing the pot. Put ½ inch of water in a pot and boil it for 10 minutes with the lid on. Then dump the water into the sink, being careful not to scald yourself with the hot steam.

Next, pour a gallon of whole milk into the pot, and warm it to about 80°F (27°C). You can do this on the stove if you have a thermometer; otherwise, just let it sit until it reaches room temperature, which will be warm enough. (You can just let the bacteria work on the milk a little longer.)

Stir in the yogurt. Cultured buttermilk will also work, and the amount is not critical; anything from a few tablespoons to a cup will do. What you are doing in this step is adding *Lactobacillus acidophilus* bacteria to the milk. These bacteria will grow and reproduce, making the milk acidic enough for the rennet to do its job. A small amount will take a little longer than a larger amount, but you are going to let this sit for 24 hours anyway, so either amount will work fine.

Put the pot (with the cover on!) in a warm place for 24 to 36 hours. You aren't making yogurt here, so the milk will still be liquid when the time is up, but it will be just the tiniest bit tart, like a weak buttermilk.

Dissolve a ¼ tablet of rennet in a few tablespoons of water. You can use a ½ tablet or a whole tablet instead if you like. Rennet is an enzyme, so it acts like a catalyst, changing the milk but not getting used up in the process. Using more makes the process go a little faster, but you are not in a hurry here—take your time.

On the stove, heat the milk to between 80°F and 86°F (about 30°C). Watch the thermometer carefully while you do this, and stir well, since it will only take a minute; you don't want to overheat it. If you don't have a thermometer, or you don't think you can do this, then don't heat the milk at all—just use twice as much rennet (½ tablet).

Stir the rennet into the milk.

Set the pot (with the lid on) in a warm place and leave it undisturbed for one to two hours. Don't rush this step. You want the milk to gel into a nice custard, so when you put your finger into it, it breaks cleanly. How long this takes will depend on the acidity of the milk, the temperature, and how much rennet you put in. It is better to just wait than to raise the temperature, increase the acidity, or increase the rennet. This is not fast food.

When the rennet has done its job and you have a nice custard consistency, it is time to slice the custard into curds.

With a knife long enough to reach the bottom of the pot, slice through the custard in ½-inch-wide strips.

In the photo below I jiggled the pot to make the cuts visible, so they look wavy. When the pot isn't moving, they are straight.

Rotate the pan 90° and repeat the slicing so you have ½-inch-square curds, like a pot full of french fries.

The next step is heating the curds so that they expel the whey. This step determines the final firmness of the cheese. I like very firm cheese, so I heat the curds to 110°F (43°C). For a softer cheese, use a lower temperature, down to as low as 102°F (39°C).

For a cheese that has firm bits of dried cherries in it, you want a nice solid cheese, so that it cuts nicely with a sharp knife and doesn't fall apart as soon as the knife hits a cherry.

Stir well but gently while heating on a low flame. Watch the temperature carefully, since it is very easy to overheat at these low temperatures. Don't stir so hard that the curds get too small, but make sure that the curds from the bottom of the pot are well mixed with those from the upper parts.

When the proper temperature is reached, turn off the flame or remove the pot from the burner. The curds need to sit at this temperature for 10 to 15 minutes, so that they firm up and lose moisture. You want them to feel like well-cooked scrambled eggs. Your assistant can help pass the time as you wait for the curds to firm up and sink to the bottom of the pot.

When the curds are nice and firm and have sunk to the bottom, strain them in the strainer. If the curds have not sunk to the bottom of the pot, it means that you have grown some gas-producing bacteria along with the acid-loving ones. While it is an indication that you did not sterilize the pot well enough, it is not a problem. It just means you will be making Swiss cheese—the gas-producing bacteria will make bubbles in the final cheese.

You can rock the strainer a bit to help the whey drain out faster. You want it fairly dry for the next step.

In this step, you add 2 teaspoons of salt and stir it into the curds well. The salt is not just for flavor (although that is important; it shouldn't taste like bland cottage cheese). You will be letting your cheese ripen in the refrigerator for anything from a couple days to several months, depending on how sharp you want the taste, and the salt will prevent bacteria and molds from spoiling the cheese.

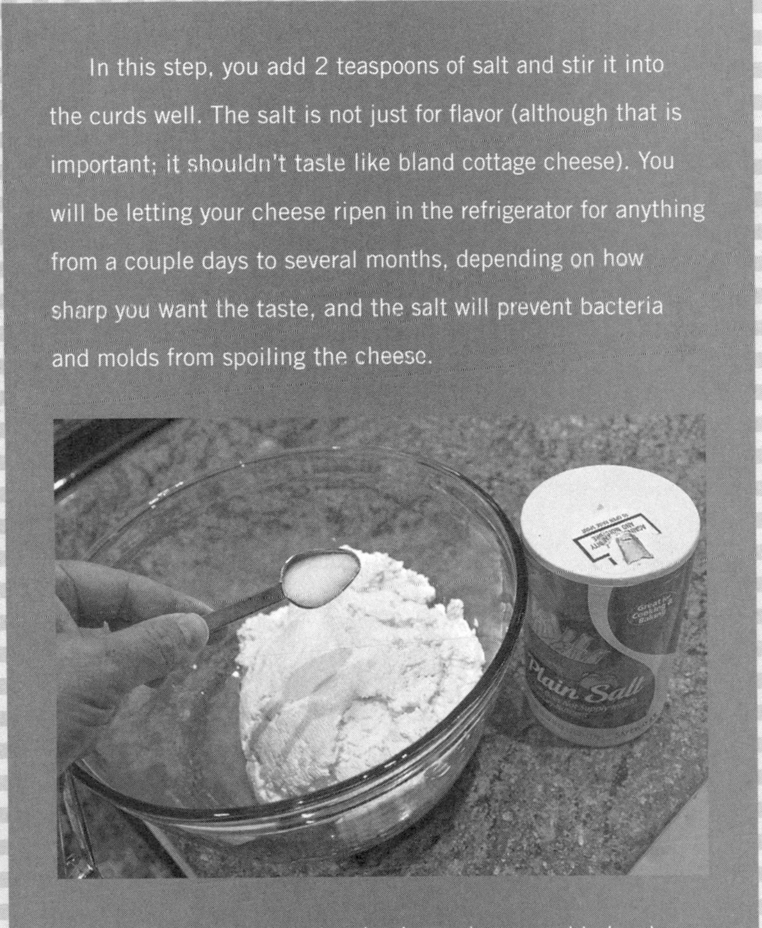

Put ½ cup of dried cherries in a microwaveable bowl, then add 1 tablespoon of sugar and 1 tablespoon of water. Microwave for a minute or two on the highest setting, so the cherries plump a little and the sugar makes a thick cherry syrup.

Let the cherries cool a bit so you don't burn your fingers, and then stir them into the salted curds. You want each cherry to be surrounded by curds, and evenly spaced though the mix.

Place the ABS pipe in a flat-bottomed bowl, and use your hand to stuff most of the cheesecloth into the pipe, leaving just a bit left out at the top. Use a rubber band to keep it in place while you fill the cheese press with the curds. In the photo you can see the disk I cut out of the plastic lid to act as the top of our makeshift homemade cheese press.

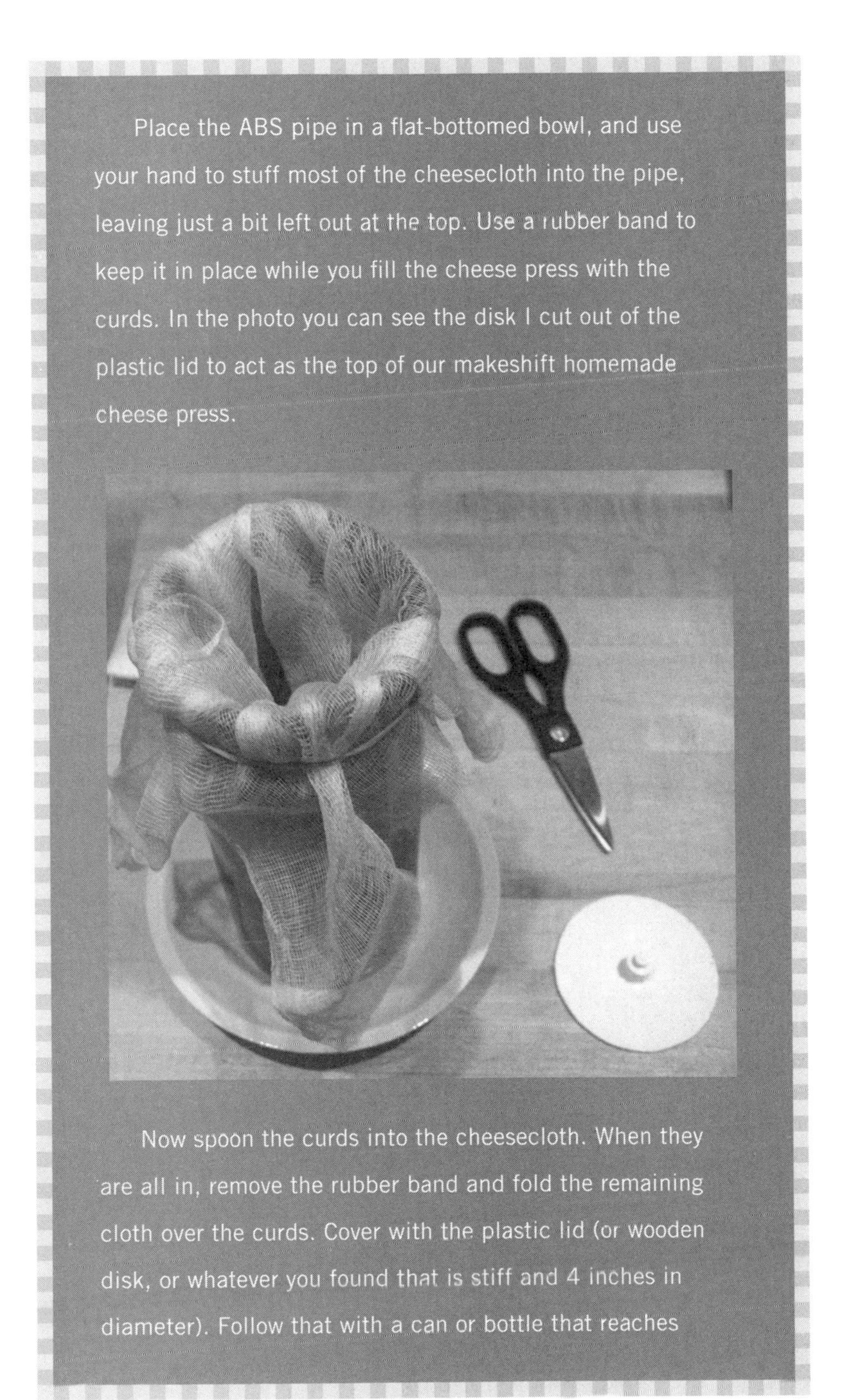

Now spoon the curds into the cheesecloth. When they are all in, remove the rubber band and fold the remaining cloth over the curds. Cover with the plastic lid (or wooden disk, or whatever you found that is stiff and 4 inches in diameter). Follow that with a can or bottle that reaches

up to the top of the press, so you can press down on it with weights.

In the photo, I used a tall plastic jar, and then inserted the handle of a dumbbell set made of plastic-coated concrete.

Stack whatever weights you are using on top of the press. You want about 50 pounds. Whey will squeeze out the bottom of the press, and you will want to soak it up in some paper towels so it doesn't spill out.

After about 36 to 48 hours in the press, the cheese is ready to come out. You may need to run a butter knife between the mold and the cheesecloth to loosen the cheese from the mold, but if the curds were firm enough, you can probably just push it out without any trouble.

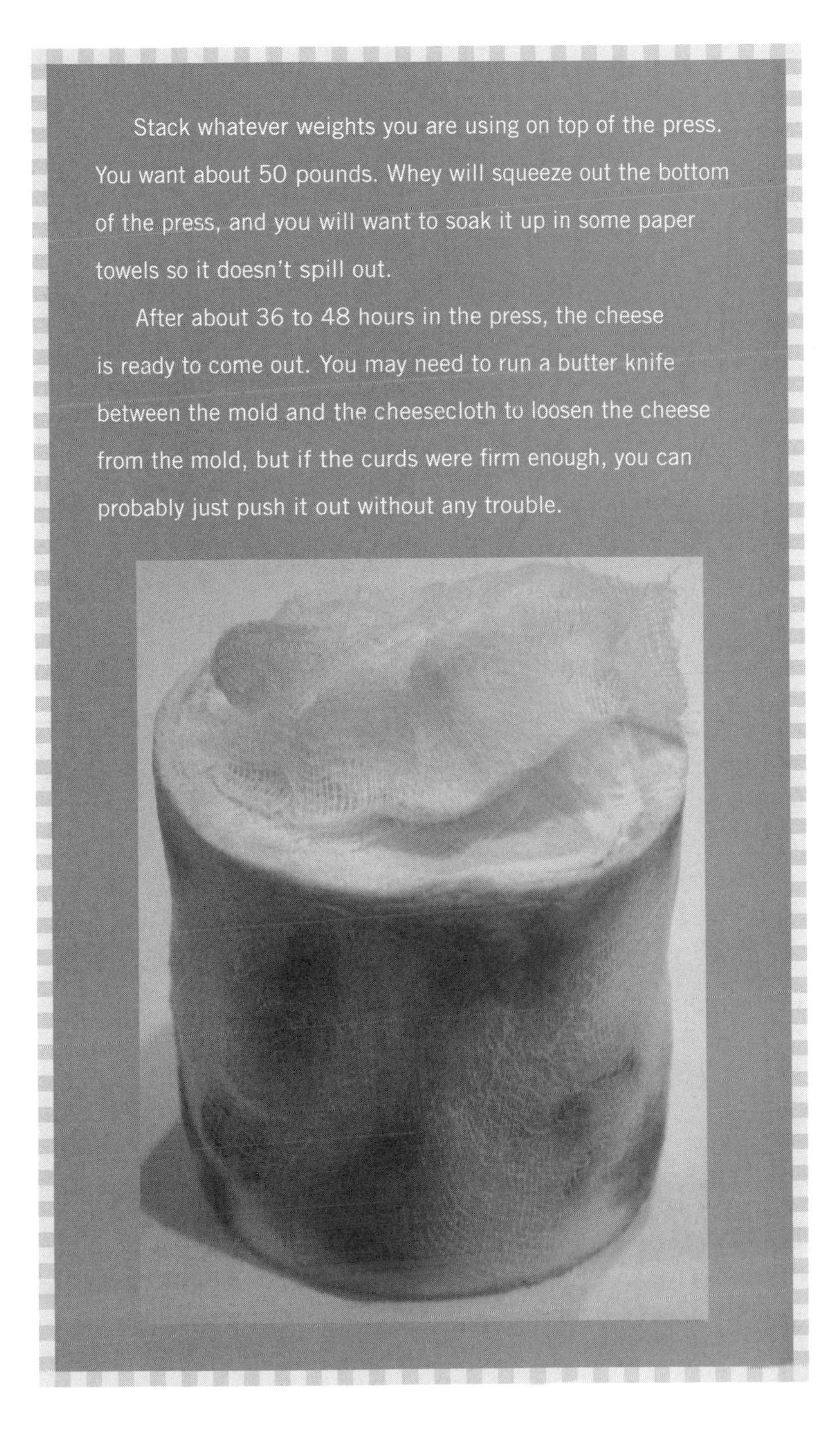

Carefully remove the cheesecloth. Some parts of the cheese cylinder will be fragile, such as the edges and any parts where cherries pressed against the cloth. Be careful on those areas so you don't pull off any cheese (just to keep it looking good).

At this point you can slice and eat the cheese. However, if you like a sharper cheese with a nice firm rind, wrap it in a paper towel or two and let it dry and ripen in the refrigerator. Replace the paper towel with a new one every day if there is any dampness on the towel. (It will probably be damp every day for the first week.)

After a couple weeks the cheese will have a nice hard rind. At this point you can wax it and let it further ripen and sharpen for months in the refrigerator.

To wax the cheese, melt a crayon and the paraffin in a disposable aluminum pie tin sitting in gently boiling water in a skillet. Stir the liquid wax to evenly distribute the color.

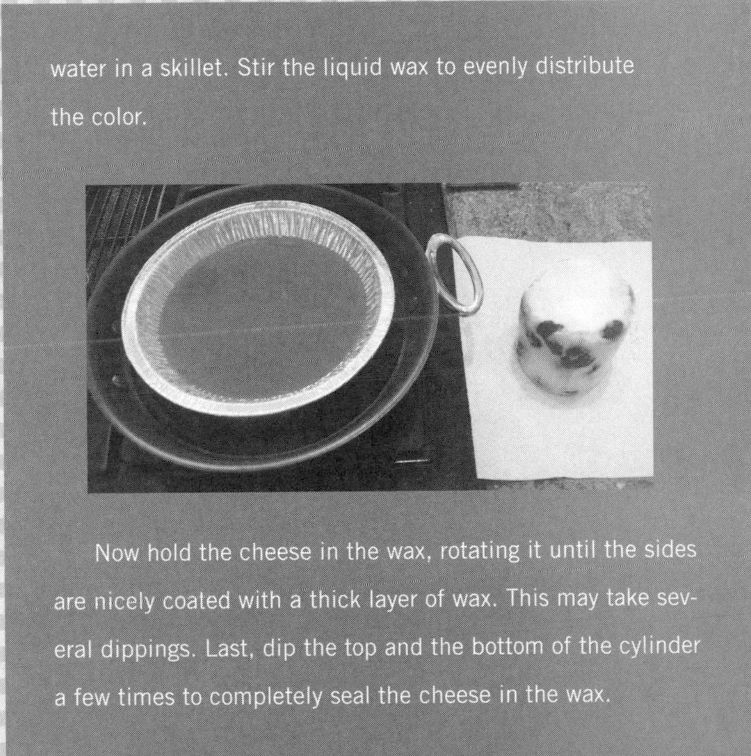

Now hold the cheese in the wax, rotating it until the sides are nicely coated with a thick layer of wax. This may take several dippings. Last, dip the top and the bottom of the cylinder a few times to completely seal the cheese in the wax.

You can label the cheese with your name, the date, and a description of the type of cheese, and the type of milk, and maybe even a suggested date for opening it some months in the future. Print this on a small piece of paper, and use hot wax to seal it onto the top of the waxed cheese.

A Holiday Variation

For the Fourth of July this year, why not create Red, White, and Blue Cheese?

Use the same recipe as the Cherry Dream Cheese, but divide the curds into three parts. In the first part, add dried Bing cherries as before.

Leave the second part alone for the white portion of the cheese.

For the third part, add dried blueberries.

When putting the cheese together in the cheese press, put the cherry cheese on the bottom. Use the disk of the press to squeeze, by hand, as much liquid as you can from the cheese and create a nice flat top.

Next comes the plain white cheese. Again, use the disk of the cheese press to squeeze the moisture out and flatten the top.

Last comes the blueberry cheese. Flatten it as before, then fold over the cheesecloth and set up the press as you did in the original recipe.

I used a 200-thread-count bamboo/cotton cloth as my cheesecloth, and it came out beautifully tie-dyed when I unwrapped the cheese. If you use a clean white T-shirt as your cheesecloth, you can wash it and wear it to the Fourth of July picnic when you serve the cheese. The colors probably won't survive many launderings, however.

The cheese can be served after a day in the press. This means you will want to start making the cheese on July 1 at the very latest. I planned to wax and age the cheese, so I was making it in mid-April. But I made two batches, one to age in the refrigerator and one to eat right away. I also made a batch on July 1 so that guests at the party could compare the fresh cheese to the aged one.

How to Make a Cheese Press

The stack of weights I used in the Cherry Dream Cheese for my makeshift cheese press worked great, but it was not the most convenient or portable cheese press in the world. So I made a new cheese press from some pine boards and bungee cords, and it works great. It also weighs 50 pounds less than the last press.

It is very simple to build. Here's what you need:

- 5-inch square or larger ¾-inch-thick pine board to cut into a 4-inch circle with the circle cutter
- 4-inch circle cutter for your drill

- 2 small finishing nails
- 9-inch length of 3½-inch-by-¾-inch pine board
- 4 eyelet screws with wood screw points
- 7-inch-by-7-inch square of ¾-inch pine or other wood
- 6-inch length of 4-inch diameter ABS pipe
- 2 bungee cords, 18 inches long

Cut a 4-inch circle of wood using the circle cutter. You can use a jigsaw instead of the circle cutter, but a circle cutter makes a nice, clean, round disk.

Nail the disc to the 9-inch board.

Screw in the four eyelets near the corners of the 7-inch square board.

Use some salad oil—canola, corn, safflower, or sunflower oil will all work fine—to waterproof all of the wood parts. Use plenty of oil, and let it soak into the wood. You can use a paper towel to spread it around, and some more to wipe off the excess.

The press can be used right away. The oil will dry and harden over the next few weeks, but it does not need to be dry to use the press.

To use the press, put the pipe in the center, on top of a pad of folded paper towels (to absorb the whey). Then insert the cheesecloth or a clean piece of an old bedsheet.

Place the curds into the cloth in the pipe, and fold the cloth over the top.

Insert the disk, and pull the bungee cords over the top of the board the disk is attached to. This will not be particularly easy, since you want something between 20 and 60 pounds of force. If you aren't strong enough to do this, cut a bit off the end of the 9-inch board until it is not too much of a bother. With less pressure, you can just let the cheese press for a longer time.

5

OILS AND FATS

The uses and properties of oils and fats are familiar to everyone who eats. But what gives them those properties is not so commonly known. You know that oil and water do not mix, but what other properties of oils and fats can you list, to better understand why they behave the way they do?

- Oils and fats float on water.
- Oils and fats have more calories per gram than sugar and protein.
- Oils are more viscous than water.
- Fats soften slowly instead of melting at one temperature like water does.
- Oils don't evaporate as quickly as water.

A water molecule is polar, with one side positively charged and the other side negatively charged, and tends to link up with other water molecules. The positive end is attracted to the negative end of another molecule, and they draw close together due to the attraction. This makes water dense.

Oils and fats are *not* polar. They don't have positive and negative parts that attract one another. Instead, they are attracted by much weaker but similar forces that arise because the electrons in the molecules are always moving. When electrons move to one side of a molecule, a temporary state exists where one side is more negative than the other. Electrons in a nearby molecule are attracted to the more positive side, and for a very short time the two molecules are synchronized, with their electrons on the same side, so they behave as if they were polar. But this attraction is very weak; small amounts of heat energy break these links, and they form and break continuously.

Oils and fat molecules are only weakly attracted to one another and do not pack together as tightly as water molecules do. So oil is less dense than water and floats.

You get energy from food by "burning" it—combining it with oxygen. If you look at a sugar molecule, you see that it already contains oxygen. In other words, it is already partially burned:

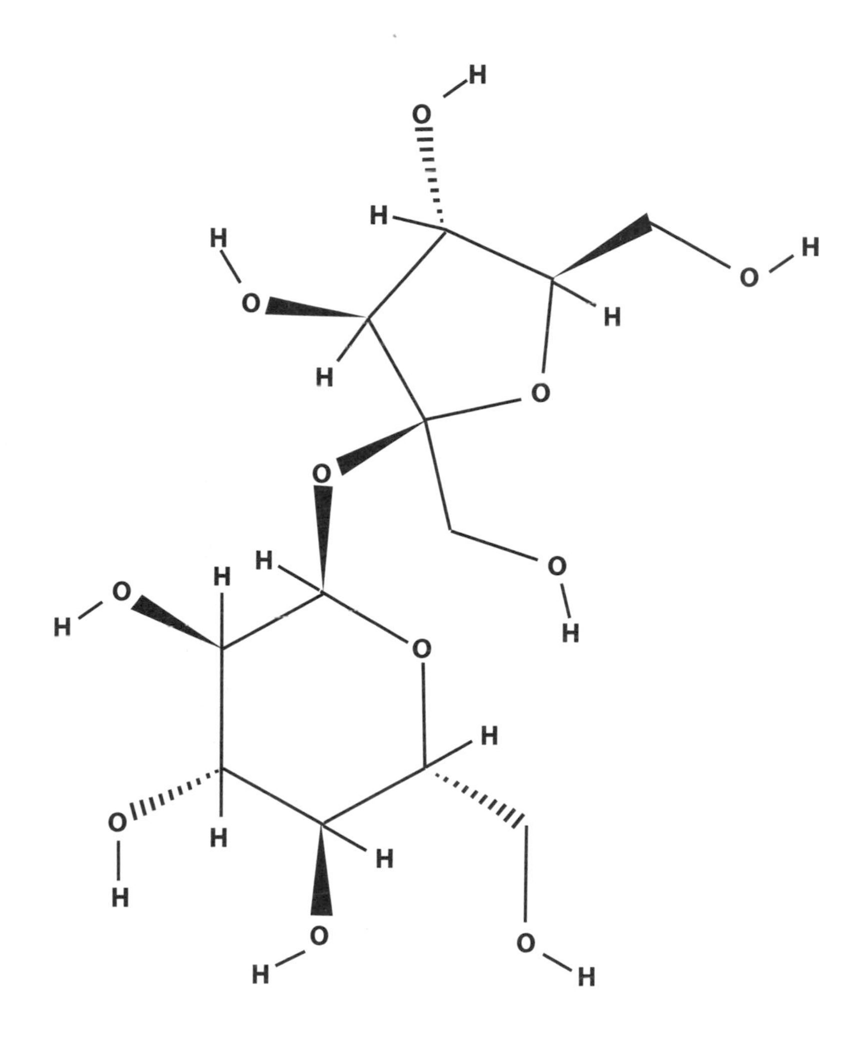

In a sucrose molecule (table sugar) there are 11 oxygen atoms, 22 hydrogen atoms, and 12 carbon atoms. Because there are twice as many hydrogen atoms as oxygen atoms, and H_2O is water, these molecules (sugars, starches, pectins, etc.) are called carbohydrates—carbon plus water.

Chemistry Lesson

Different Ways to Look at Molecules

In this book I have chosen to show molecules using the most common shorthand form. But there are other ways chemists use to show the structure of a molecule.

The following illustrations are different ways of showing what oleic acid looks like. Oleic acid is one of the fatty acids in olive oil.

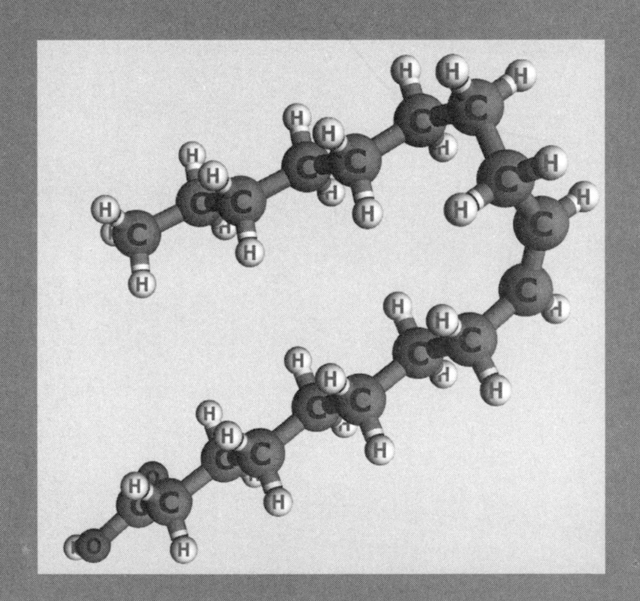

Here you can see the carbons, hydrogens, and oxygens labeled. All of the labels make the picture very busy.

When a molecule contains only a few different types of atoms, we can simply use color or shading to identify them.

The space-filling form above is less cluttered, but some atoms are almost hidden. The following form shows them more clearly.

Here you can see where the double bond is, and that there is only a single hydrogen attached to each double-bonded carbon.

Fat molecules are mostly just carbon and hydrogen. A typical triglyceride (fat) only has 6 oxygen atoms and might have over 50 carbons and over 100 hydrogens. Because less of the molecule is already burned, fats contain more fuel than carbohydrates.

A typical triglyceride found in fats is tristearin, found in beef fat:

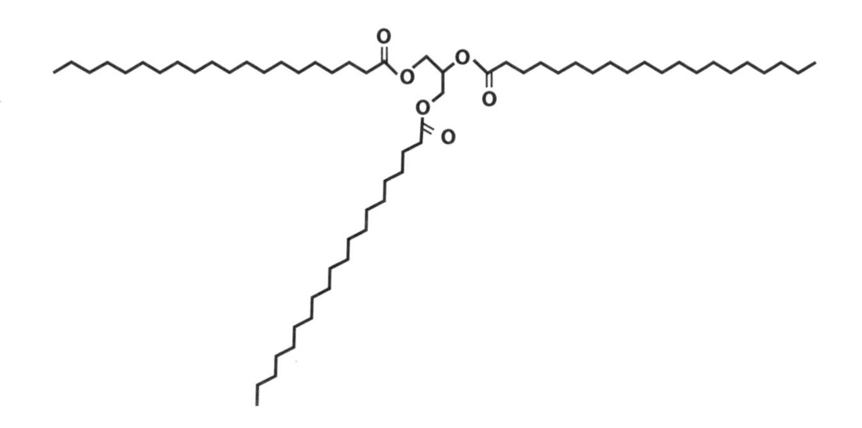

It has a glycerin molecule

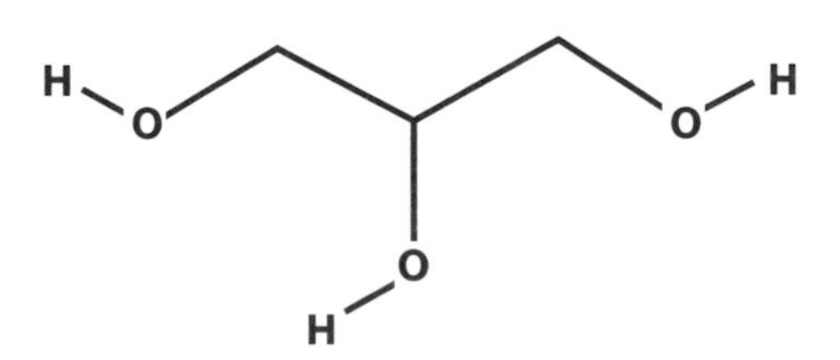

at its center, and instead of the three hydrogens attached to the oxygens, it has three stearic acid molecules attached:

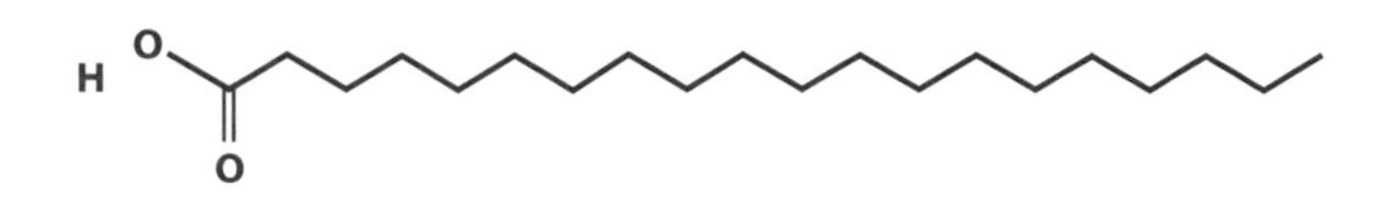

Three long chains of carbon and hydrogen are attached to the central glycerin molecule. These long chains rub against and tangle with other long chains on nearby tristearin molecules, which makes it more difficult for the molecules to move about than is the case for water molecules. The result is a thicker liquid, with a higher melting point than water. In fact, tristearin is a solid at room temperature, melting at 161°F (72°C). But it is still less dense than water, and it floats.

If you had a pile of branches from a tree, the branches would not pack flat, whereas in a pile of lumber all the boards will line up and pack flat, with each board touching the next all along its length. The long chains in fats like tristearin can line up together, like lumber, so the attractions between the molecules are stronger than if they only touched at one or two points, like branches. This attraction raises the melting point. There are triglyceride molecules where the long chains have kinks in them, like the branches, so they don't line up easily. (See the kink at the double bond in the oleic acid molecule on the bottom of page 83.) These have lower melting points.

Oils have more of these kinked molecules than fats do, so they are liquid at room temperature. The kink forms at the double bond, because the lone hydrogens attached there are on the same side, have a slight positive charge, and repel one another. The slight positive charge is because the electron is attracted to the carbon more than the hydrogen and therefore spends more time near the carbon than the hydrogen.

Because the molecules are so big, it takes more energy to get them to leave the liquid state and become a gas. Oils and fats do not evaporate as easily as small molecules like water.

Most fats are a mixture of many different types of triglyceride molecules. In pure form, each molecule would make a liquid that has a distinct sharp melting point. But when many different types are all mixed together, the substance softens slowly over a range of temperatures instead of melting at one temperature. When a substance such as ice (which has only one type of molecule: water) melts, it does not soften slowly. Instead, it has two distinct phases. The solid ice is coated with liquid water, and the two are at different temperatures. But fats (because they are composed of many molecules with different melting points) slowly become soft and can be spread or mixed with other ingredients before they are completely melted.

Saturated Fats

Carbon atoms can bond to four other atoms. In the long chains in the triglyceride molecules, each carbon is attached to the next carbon atom by either a single bond or a double bond. A double bond uses up two of the four available bonds, so each double-bonded carbon can only hold onto one hydrogen instead of two. If the carbons are attached with only a single bond, each carbon has room to hold onto two hydrogen atoms. Fats that have as much hydrogen as possible are said to be saturated with hydrogen; they are saturated fats.

The tristearin in beef fat is an example of a saturated fat. Saturated fats have chains that are not kinked. They can lie together closely like a stack of lumber.

Monounsaturated Fats

A double bond in a chain of carbons introduces a kink in the molecule. Because there are two bonds, the molecule is constrained in its motion, like a door with two hinges. The kink makes it harder for the molecules to line up together, so the attractive forces are less. Where a saturated fat has chains that are flexible, like a necklace chain, a chain made of all double bonds would be like a bicycle chain, only able to bend in one plane.

When there is only one double bond in the long chain, the fat is said to be monounsaturated. A typical monounsaturated fat is triolein (found in olive oil and macadamia nuts). It stays liquid at temperatures below freezing (22°F, -5°C). The oil in macadamia nuts is almost 80 percent monounsaturated, and olive oil is 75 percent monounsaturated.

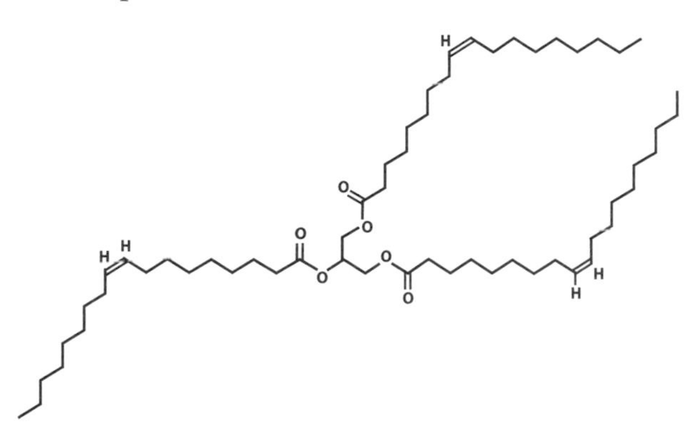

Triolein

Polyunsaturated Fats

When more than one bond in the chain is double, it's a polyunsaturated fat. As the number of double bonds increases, the melting point of the fat begins to go up again. But unlike saturated fats, where the high melting point is caused by the straight chains stacking up next to one another, the double bonds of polyunsaturated fats get tangled together, which also raises the melting point.

Chemistry Lesson

Kinky Molecules

Sometimes a concept is easier to understand if you can feel it in your hands. To get a better sense of how a kink forms at a double bond, look at the photo below. It shows eight cubic magnets and nine steel balls. The magnets represent carbon atoms, and the steel balls represent the clouds of electric force where the electrons are.

The two steel balls side by side on the left represent a double bond. It is easy to bend the molecule at this point, but only along a single plane, like a door on a hinge. The rest of the chain stays somewhat straight, since the negative charges (steel balls) repel one another and try to stay as far apart as possible. But they can still rotate.

An example of a polyunsaturated fat has the conveniently memorable name of trieicosapentaenoin (triEPA for short). In triEPA, each of the three chains has five double bonds. The name comes from the fact that there are 20 carbons in the chain (the Greek word for 20 is *eicosa*), and five double bonds (the Greek word for 5 is *penta*).

Trieicosapentaenoin is found in fish oils. The fish don't make it; they get it from the algae they eat. The three chains are made up of eicosapentaenoic acid (EPA for short), which the human body converts into many essential substances called eicosanoids.

EPA is also converted into docosahexanoic acid (DHA for short), the major fatty acid in sperm and the precursor molecule to many important hormones.

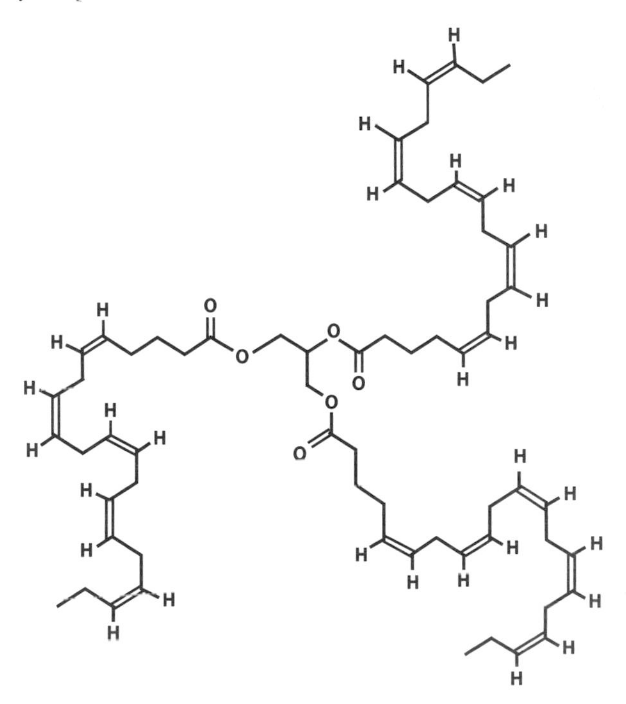

Trieicosapentaenoin

Omega-3 and Omega-6 Fats

If you look at the ends of the three chains in triEPA, you see that the last carbon in the last double bond is attached to a very short (two-carbon) chain at the end. The tail of the chain is called the omega end, named for the last letter in the Greek alphabet. The third carbon from the end is the one with the double bond.

Such a fat is called an omega-3 fat. More commonly, each of the three chains is said to consist of an omega-3 fatty acid.

In humans, omega-3 and omega-6 fatty acids are used to create many hormones that control the immune system and other functions. The enzymes that perform the conversions are the same for both types of fatty acids. But the omega-6 fatty acids can hog these enzymes, so that more omega-6 hormones are produced than omega-3 hormones. This can cause inflammation, arthritis, heart attacks, strokes, and depression. The typical Western diet contains 10 times as much omega-6 fat as omega-3 fat, resulting in hormonal imbalances that contribute to all of those diseases.

An example of an omega-6 fat is triarachidonin:

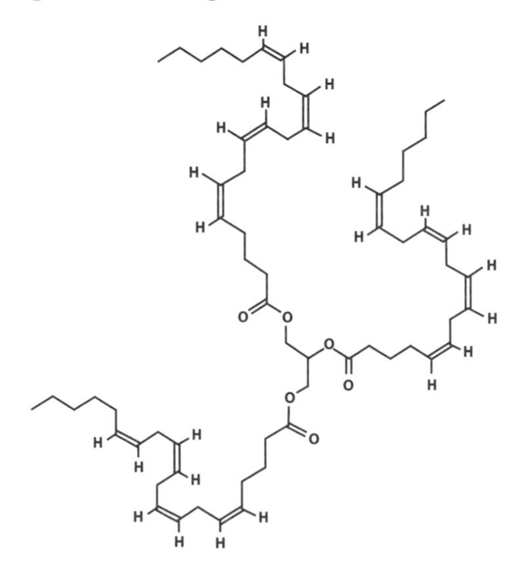

The drugs called COX-1 and COX-2 inhibitors used to treat pain and inflammation (aspirin, ibuprofen, naproxen) work by preventing arachidonic acid from being converted into inflammatory compounds.

Trans Fats

The fats discussed so far have both hydrogens at each double bond on the same side of the chain. The hydrogens are said to be in the *cis position*, from the Greek word for "on the same side." When the hydrogens are both on the same side, the chain kinks.

But there is another position they can be in, in which one hydrogen is on one side and the other is on the opposite side. This is called the *trans position.*

The chains in trans fats are straight, like those of saturated fats. Trans fats have high melting points and act a lot like saturated fats.

Trans fats form when unsaturated fats are heated. As the heat makes the molecules shake and vibrate, some of the bonds get changed from the cis to the trans position. As the fat continues to cook, the bonds may change back and forth many times, resulting in a fat that has about half of the bonds in the cis and half in the trans positions.

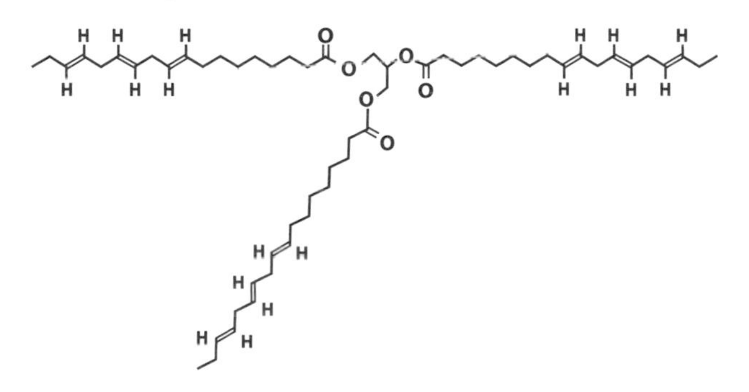

Trans trilinolenin

In the making of vegetable shortening, vegetable oils containing unsaturated fats are heated with a catalyst and some hydrogen, so that some of the double bonds gain a hydrogen and convert to saturated single bonds. But the process requires heat, and the heating produces undesirable trans bonds.

Trans fats have been shown to increase the risk of coronary heart disease by raising the levels of bad LDL cholesterol and lowering the levels of good HDL cholesterol.

Shortening was invented as a way to make solid fat from cheap vegetable oils, which could be substituted for the more expensive lard in cooking. It was only recently that the health problems associated with trans fats were discovered. Originally, vegetable shortening and margarine were sold as healthier alternatives to lard and butter.

These days, shortening is made by continuing to saturate the oils, so that there are almost no double bonds left, and then mixing the product with unsaturated oils until there is less than a gram of trans fat in a tablespoon of shortening. This allows the manufacturer to claim that there are zero grams of trans fat in the shortening.

Since there are 12.8 grams in a tablespoon of shortening, there can be 6 percent trans fat left in the shortening to still qualify for the label "zero grams" of trans fat. In other words, there can be 16 grams of trans fat in a cup of shortening that claims to have zero grams.

Compare the structure of triolein shown earlier with its trans version shown on the next page:

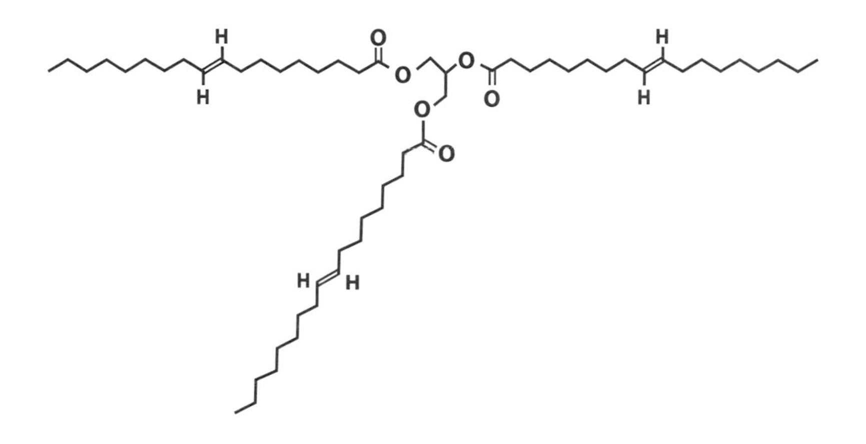

Here the chains are not kinked. They are shown as being straight, but actually the long chains can bend like spaghetti, except where there are double bonds. There, the movement is constrained to a hingelike motion, due to the carbons being linked in two places instead of just one.

Saturated fats have the most flexible chains, since all the bonds are single bonds. They can pack together tightly in many configurations, and so they have the highest melting points. Trans fats are almost as good at packing, and have melting points in between that of similar saturated and cis-unsaturated fats. In the kinked cis-unsaturated fats like triolein, the kinks get in the way of packing, and the melting point is considerably lower.

6

SOLUTIONS

Solutions are just mixtures in which two or more substances are well mixed (homogeneous). We are used to thinking of liquid solutions for which a solid substance (the solute) is dissolved in a liquid substance (the solvent). But liquids can be dissolved in other liquids, and solids can be dissolved in solids, and gases can be dissolved in liquids.

Some questions this section will attempt to answer are:

- Why do things dissolve?
- Why do different liquids dissolve different solids?
- Why do heat and stirring make things dissolve faster?
- What limits how much of something will dissolve?

To begin, look at table salt dissolving in water. Water is a polar molecule, with one side slightly positive and one side slightly negative.

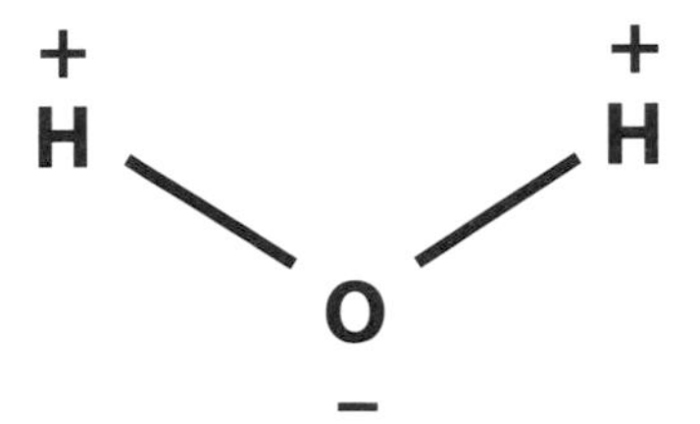

Ionic bonds were discussed earlier. An ion is formed when an atom loses an electron, or gains one, and becomes charged. Table salt is an ionic solid composed of positively charged sodium ions and negatively charged chlorine ions. Their opposite charges make them stick together closely, making the result a dense solid. The ions are created when sodium atoms lose electrons to chlorine atoms.

When solid table salt is dropped into water, the positive ends of the water molecules are attracted to the negative chloride ions, and the negative ends of the water molecules are attracted to the positive sodium ions. The attraction goes both ways—the electrons in the chloride ions move closer to the water molecules, and the electrons in the water molecules move closer to the sodium ions. This makes the attraction between the sodium ions and the chlorine ions weaker. Several water molecules exert a stronger attraction to the ions than the ions have for one another, and the ions move into the water. The salt has dissolved.

The reaction is reversible. Ions in the water attract one another when there are not enough water molecules to overcome the attraction, and salt crystals grow. As the area next to a dissolving crystal of salt becomes saturated with ions, the rate of dissolving matches the rate of crystal growth. Things would come to a standstill if it weren't for the motion of the water molecules. That motion carries the ions farther away and allows more water molecules to get close to the salt, so the dissolving process continues.

When the solution eventually becomes so full of ions everywhere that the rate of crystal growth matches the rate of dissolving, the solution is said to be *saturated.* The saturation point is affected by the heat of the solution. Heat is just the motion of molecules, so when the solution is hotter, the molecules jostle around more. This makes it more likely that an ion will leave the crystal than that one will stick to the crystal, so more salt will dissolve in hot water than in cold water.

To make the crystals dissolve faster, stir the water. This mixes the pure water with the saturated solution close to the crystal, so there are more water molecules next to the crystal than there would have been without stirring.

A similar effect happens when sugar is dissolved in water. Sugar is not an ionic solid like salt. Instead, it is a polar molecule, like water. There are places on the molecule that are slightly more positive and places that are slightly more negative. These areas attract one another, and because there are many of them—the molecule itself is somewhat large—sugar is a solid.

In water, the positive and negative parts of a sugar molecule attract the negative and positive parts of the water. Again, the attraction of a lot of water molecules is stronger than the attraction of the sugar molecules to one another. The sugar dissolves.

It takes energy to break the bonds that make salt or sugar solid. The energy can come from many sources. One is the heat of the water and the salt itself. Another is the attraction of the water molecules to the molecules of the salt or sugar. Just as the gravitational attraction between the earth and the water behind a dam can provide energy for generating electricity, the attraction between the polar water molecules and the polar molecules in the solid can weaken the strength of the bonds in the solid.

Something like sand does not dissolve in water. The molecules in sand are bound so tightly together that their attraction to one another is stronger than their attraction to water, so nothing happens.

When a nonpolar molecule like Styrofoam dissolves in a nonpolar solvent such as gasoline, similar processes take place. The bonds involved are generally weaker, and diffusion becomes the most apparent force. Diffusion is the tendency of things that are separate to become mixed up when randomly shaken by the motion of molecules (heat). If you drop a bit of food coloring into a glass of water you can watch it slowly diffuse until the entire liquid is a uniform color.

In the same way, the Styrofoam molecules gradually diffuse into the gasoline until the two are completely mixed. The same effect happens when salt or sugar dissolves in water, of course, but without water's polar nature, the relatively strong bonds between the salt molecules or the sugar molecules would prevent diffusion from happening. This is why salt and sugar do not dissolve in oil.

Higher temperatures generally increase solubility, but the effect differs for different substances. At just above freezing, about 356 grams of salt will dissolve in a liter of water. At just below boiling, 390 grams will dissolve, which is not a big increase. For sugar, however, the amounts go from 1,790 grams at near freezing to 4,870 at near boiling.

Adding other substances can affect solubility. After you have dissolved as much sugar as possible into water, adding a little salt will allow more to dissolve. In effect, some of the sugar is dissolving in some of the salt, and vice versa. The various salts in mineral water can thus make a difference in candy making.

It is possible to supersaturate a solution. You can dissolve as much sugar as possible into hot water and then carefully cool it without stirring, and the sugar will remain dissolved. Drop a few crystals of sugar into the solution at that point, and they will grow into large sugar crystals (rock candy) as the sugar crystallizes out of the solution. The sugar molecules are more strongly attracted to one another than to the smaller number of water molecules available in a supersaturated solution.

Seltzer and Temperature

Gases also dissolve in liquids. Carbon dioxide dissolves in water to make carbonated water. Here, temperature has the opposite effect. Since carbon dioxide is a gas, there are no bonds holding it together that have to first be broken before it can dissolve. But the bonds between the gas and the liquid break more easily with increased temperature. Thus more carbon dioxide will dissolve in cold water than in warm or hot water.

The nitrous oxide in whipped cream cans, or "whippits," exhibits the same effect. More of it will go into solution in very cold cream than in warm or room temperature cream.

When making ice, you sometimes want to prevent bubbles of air from forming in the ice so it stays clear. As water freezes, crystals of solid ice form. Crystals are, in general, very pure—as layers of molecules accumulate in the crystal, foreign molecules are left in the remaining liquid. Eventually, they exceed the saturation point and come out of solution. If the foreign material is dissolved air, bubbles in the ice are the result.

To prevent the bubbles, boil the water before freezing it. The high temperature makes most of the air come out of solution, so there is little left in the water when it is subsequently chilled. Since air can dissolve back into the water as it cools, cover the water with a sheet of plastic wrap to prevent contact with the air.

Syrups, Broths, and Other Solutions

There are many solutions in the kitchen. Beef broth is a solution of salt, proteins, sugars, and other small molecules. Sodas are solutions of sugar, flavorings, and carbon dioxide. Vinegar is a solution of acetic acid and small flavor molecules in water.

In the previous discussions about colloids, sauces were thickened using small amounts of proteins, starches, or other large molecules. But because you can dissolve almost four and a half pounds of sugar in a quart of water, even a small molecule like sugar will thicken water if there is enough of it. This is no surprise—at those ratios, there are only about 10 molecules of water per molecule of sugar. In boiling water, where you can dissolve even more sugar, there are only about four molecules of water per molecule of sugar.

The density of maple syrup, honey, and corn syrup is about 1.3 times that of water. Thus a liter of syrup has about 300 grams of sugar in it, about 10 ounces per quart.

The sugar in corn syrup is glucose. It is made by using enzymes to break down cornstarch, which is made up of chains of glucose molecules. Glucose is not as sweet as sucrose, so corn syrup does not taste as sweet as sugar syrup does. Adding to the viscosity are long chains of glucose molecules that were not completely broken apart by the enzymes used to change starch into glucose.

High-fructose corn syrup is made by using other enzymes to convert some of the glucose into the sweeter simple sugar fructose. To get a syrup that has the same sweetness as sucrose syrup, a ratio of 55 percent fructose to 45 percent glucose is used. This is called HFCS 55. This is the same ratio of fructose to glucose that honey has. Because US corn is subsidized and sugar has tariffs and quotas, high-fructose corn syrup is less expensive than sugar. But for the most part, it is flavorless honey.

Honey itself is a supersaturated solution of fructose, glucose, maltose, and sucrose. The water and sugars make up 99.5 percent of the weight. The other 0.5 percent is a mix of protein, amino acids, vitamins, and minerals, although the amounts are minute in comparison to the recommended daily allowances of those nutrients. Honey is between 25 percent and 44 percent fructose and between 24 percent and 36 percent glucose, with the other sugars making up less than 9 percent by weight.

Sucrose syrup can be heated with a small amount of acid to separate the fructose and glucose. The result is invert syrup, which is slightly sweeter than the original sucrose syrup. It is also less likely to crystallize. Both of these properties come into play when making jams and jellies, where the fruit adds the acid needed.

Candy

Solutions are not always liquid. Many hard candies are solid solutions of sugar, water, and flavorings.

Rock candy, where sugar crystals are carefully grown to large size, is not a solution. But lollipops, suckers, and other clear, hard sugar candies are a form of glass made from sugar.

In making hard candies, several tricks are used to ensure that the sugar does not crystallize into an opaque, gritty texture. Corn syrup is used along with the sugar because the unbroken chains of glucose left over from converting starch help to prevent crystallization. Acids such as tartaric acid or citric acid are used to convert the sucrose into fructose and glucose. Having two dissimilar simple sugars in the solution also helps to prevent crystallization, which occurs best with pure substances.

The water, sugar, corn syrup, and acids are cooked together until the amount of water falls to a very low level. When the solution comes to a boil, no further stirring or agitation is done, to prevent the fructose and glucose from joining back together into sucrose. Any crystals of sucrose are washed from the walls of the pan to prevent them from acting as "seed" crystals, and making the whole batch of candy suddenly crystallize into a gritty or sandy texture.

Flavorings boil away if added too soon, so generally once the mixture has been cooked to the hard-crack stage (310°F, 154°C), it is cooled to 275°F (135°C) before the volatile flavorings such as vanilla or mint oils are stirred in.

The mixture is then cooled quickly to prevent crystallization, often on a marble slab or a baking pan turned upside down to speed the cooling. Fast cooling freezes the molecules in place, giving them no time to rearrange themselves into precise crystalline order.

Liquors

Liquids can also dissolve in other liquids. Some liquids behave like any of the solids previously discussed when dissolving. They will only dissolve a certain amount and then the solution saturates. No more will dissolve, leaving the liquids in two distinct phases.

Other liquids, such as ethanol and water, will dissolve in one another in any proportions. They are said to be completely *miscible* in one another. This feature of ethanol and water allows alcoholic beverage makers the entire range of potencies—from less than 1 percent alcohol content to 100 percent alcohol.

Many flavor elements are nonpolar oils and fats. Because they are nonpolar, they dissolve in alcohol better than they do in water. But if there is not enough alcohol (too much water), they come out of solution. Absinthe becomes cloudy, or oils begin to float on the surface of the liquid. Sometimes this is beneficial, bringing the aromatic oils closer to the nose as ice melts in a glass.

The opposite effect happens with sugar, which dissolves better in water than in alcohol. Having both solvents in the same glass allows a sweet drink to contain aromatic oils and other flavors that do not dissolve in water.

7

CRYSTALLIZATION

Crystallization is important in many aspects of food preparation and storage as well as in creating and refining the ingredients for foods and household products.

Crystallization is used to purify sugars and fats. It is used to change the texture of foods such as ice cream, fondant, fudge, and chocolate. Controlling crystallization is important when freezing foods.

The difference in the way dark chocolate snaps when you break it from the way milk chocolate breaks is due to differences in the way the cocoa butter is crystallized in the two. Chocolate is "tempered" by controlling the heat while it cools, in order to make a lot of very small crystals. The white bloom that sometimes forms on chocolate that has been stored improperly is a result of changes in the crystals of the cocoa butter from one type of crystal to another.

Ice cream is smooth because its billion or so crystals of ice per liter are, on average, only 40 microns across. Over time (if the ice cream in your house ever lasts long enough in the freezer) the ice crystals grow, and the result is a grainy texture.

When a crystal forms, molecules of a particular compound arrange themselves in a close periodic order. Molecules of different compounds are either too large or too small, or do not bind to the others in the same way, and are excluded from the crystal. Crystals are thus pure substances. Allowing something to crystallize is a way of purifying it.

Pure substances melt at a definite temperature. Solids that are not pure crystals, but instead a mixture of different crystals, gradually soften over a range of temperatures instead of melting all at once. By watching the temperature of a substance as it is heated, you can tell whether it is a pure substance or a mixture. If there is a sharp melting point, you know it is a pure substance.

Pure cocoa butter melts in a very narrow range between 97°F and 99°F (36°C and 37°C), right at the temperature in your mouth. In your hands, which are colder than the inside of your mouth, dark chocolate (where the fat is pure cocoa butter) does not melt. Contrast this with milk chocolate, which contains a mixture of different fats (including butterfat) and softens over a wider range, so you get sticky fingers.

Cocoa butter contains only three main types of fat (in this case, monounsaturated triglycerides), while butterfat from milk contains over 400 types. Cocoa butter thus melts at a much sharper temperature point than butterfat does. Knowledge of this can help you when designing recipes for which you might want to control the melting point or spreadability of your creation. Cocoa butter is not spreadable, but butterfat is, so adding butterfat will make the cocoa butter easier to spread on toast or a graham cracker.

Sugar Crystals

Sugar is refined from cane syrup by growing large crystals. The mix of crystals and remaining liquid is separated by spinning the liquid away in a centrifuge, leaving only the crystals. Large crystals do not trap the liquid between them like small crystals do, because large crystals have less surface area to hold the water than small crystals do.

Your tongue and palate can detect graininess of sugar crystals as small as 15 microns. For a fondant, caramel, or fudge to feel smooth and creamy, the sugar crystals must be at least that small.

Controlling the Size of Crystals

In many foods such as ice cream, chocolate, caramel, fondant, fudge, and even butter and margarine, the size of the crystals is very important to the texture and consistency of the result. Whether the crystals are fats, ice, or sugar, controlling their growth is the key to the recipe.

With ice cream, you must stir the mixture constantly as it freezes. The ice crystals form at the edge of the container and are scraped off and mixed into the more fluid center. The growth of the crystals is affected by the rate of cooling and the proteins in the milk, which compete with water molecules to surround the growing ice crystal. In a similar fashion, the lactose crystals in the ice cream are also limited by proteins that coat them and prevent them from joining other nearby lactose crystals. The crystals of sucrose and lactose and butterfat also help keep the ice crystals from joining together into larger crystals detectable by the tongue.

Since the speed of the freezing process affects the crystal size the most, making ice cream with liquid nitrogen is an excellent way to keep the crystals small. One of my favorite recipes (because it is so simple) is short enough to fit into a Twitter post: "One gallon half-and-half, 2 cups sugar, 4 tbs vanilla. Add 1 gallon liquid nitrogen slowly while stirring. Serves 32." I stir it with a portable electric drill with a paint-mixing attachment. The result is very smooth and creamy. (My first attempt used heavy whipping cream and was so disgustingly rich a single scoop was almost too much.)

In chocolate making, the process of tempering the chocolate creates small crystals. The rate of crystal formation is controlled by the rate of cooling. Fast cooling makes small crystals. Slow cooling allows larger crystals to form. If the crystals are too large, the chocolate is dull and grainy, without the smooth shine of a nice dark chocolate bar. The tempering also controls whether the crystals are in one form, the β(V) form, or in another, the β(VI) form. The first form produces the desired shiny surface, and the second form causes the whitish "fat bloom" you see on chocolate bars that have been stored at too high a temperature. Fat bloom in some recipes is prevented by adding butterfat, which prevents the recrystallization by adding many different types of triglycerides to the mix.

Sometimes, however, you want big crystals. The many varieties of salt on the market are indistinguishable when they are dissolved in water. The main difference is the size of the crystals and how the crystals are stuck together. Rock candy is just huge sugar crystals (sometimes with color and flavorings added, which lodge in the imperfections of the crystals).

8

PROTEIN CHEMISTRY

Proteins have many effects on food and cooking, from stabilizing foams and emulsions to firming up gels. Knowing a little bit more detail about proteins will help you design and modify recipes and fix things that go wrong.

Amino Acids

Proteins are made up of amino acids. An amino acid has a central carbon, with a carboxyl group (COOH) at one end and an amine group (NH_2) at the other end. The simplest amino acid is glycine.

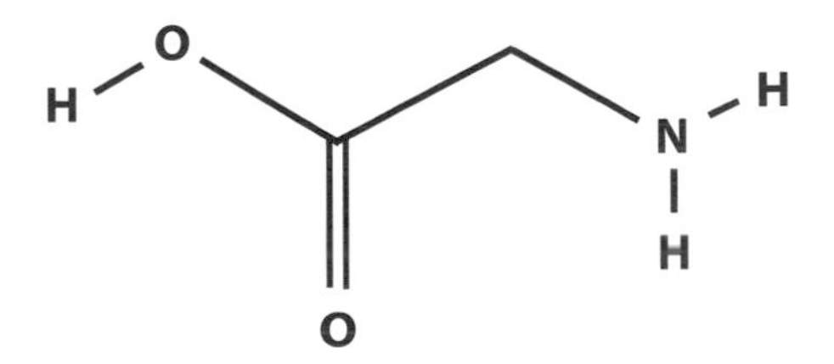

Notice that at the left end, there is a hydrogen attached to an oxygen. At the right end there is a hydrogen. Amino acids can join together by joining end to end and losing a water molecule—the OH at the left and the H at the right.

For example, two glycines can join to form diglycine. In the first drawing below, the box shows where the two glycines will join together.

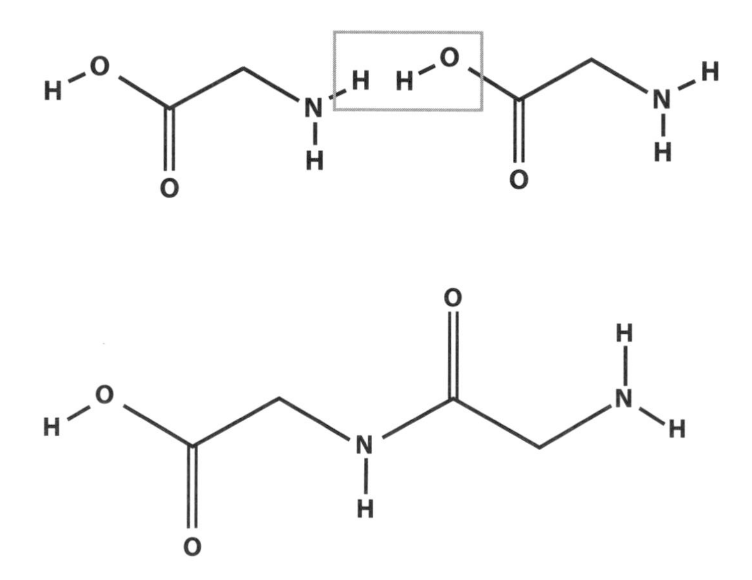

This kind of bond is called a *peptide bond*, and it is very strong. When only a few amino acids are joined together in this way, the molecule is called a polypeptide. Longer polypeptides are called proteins.

There are about 22 different amino acids found in the proteins that make up our bodies and the foods we eat. These amino acids differ from one another in one particular way: the things that are attached to the carbon right next to the nitrogen. In glycine, there is just a hydrogen atom there.

Chemistry Lesson

Four Kinds of Protein Structure

The sequence of amino acids in a protein gives it what we call the *primary structure* of the protein. You can think of the primary structure as a string of beads, in which each bead is an amino acid.

In some amino acids, long chains of atoms are attached to the carbon atom next to the nitrogen. These chains can form their own bonds with one another, to create what is called the *secondary structure*. Two common secondary structures are the alpha helix, in which the protein forms a coiled spring, or helix, and the beta sheet, in which the strings of beads bond parallel to one another to form sheets. These forms are held together with hydrogen bonds (see page 39).

The *tertiary structure* of a protein is the form it takes when it folds into a three-dimensional shape. Soluble proteins usually have a globular or almost spherical tertiary structure. As mentioned earlier, egg albumin is globular; there is also a whole class of proteins that are called globulins because of their shape. Insoluble proteins—such as collagen in connective tissue, elastin in tendons and arteries, and keratins in hair, hooves, and nails—have a fibrous tertiary structure.

Some proteins combine with other molecules to form *conjugated proteins*. In the nuclei of cells, for example, proteins combine with nucleic acids to form nucleoproteins. Proteins can combine with just a little carbohydrate to form a glycoprotein, or with more than about 4 percent carbohydrate to create mucoproteins. Combined with fat, they are lipoproteins. The structural properties of these combinations result in the *quaternary structure* of a protein—the last of the four kinds of protein structure

Denaturing Proteins

In their natural state, proteins like egg albumin and milk casein are soluble in water. Most of their hydrogen bond–forming parts are tucked inside the folded structure of the protein, making them unavailable for forming bonds with other molecules. They are all the same shape, so they all have the same properties and can form crystals.

There are several mechanisms that destroy these properties. Heat, acids, strong alkalis, alcohol, urea, salicylate, and ultraviolet light are among the more common ways that proteins become denatured.

A denatured protein unfolds as many of the hydrogen bonds that preserve the three-dimensional structure of the protein are broken. Instead of a uniform solution of molecules that are all the same shape, in a denatured protein, the molecules can take a staggering number of different shapes (on the order of 1,020 different shapes, depending on the size of the protein molecule). Like snowflakes, few if any of the molecules will have the same shape, and they will no longer form regular crystals.

The unfolded molecules also have more bond-forming areas exposed on the outside, so they form bonds with one another and coagulate. They become insoluble in water.

You have seen that surface effects cause proteins to denature. When you beat egg whites or whip cream, the proteins unfold as their hydrophobic parts rearrange to avoid water in favor of air or fat. The unfolded proteins can then bond to one another to create stabilizing protein films that keep the new forms in the desired shape.

In cooking, you control the denaturing of proteins in several ways. You can control the temperature, you can control the acidity, you can use copper bowls to beat the egg white and catalyze the formation of disulfide bonds in the proteins, and you can control the fat or air content when you beat the proteins.

For example, when beating egg whites, it is important to keep fats out of the egg whites. A little bit of oil or egg yolk can prevent the foam from stabilizing, as the air competes with fat for the hydrophobic parts of the molecule.

Because proteins have both acidic parts and basic parts, the acidity of the solution they are in changes their behavior. Acids release protons (hydrogen nuclei) and bases accept protons. In an acidic solution, the basic parts of the protein accept protons from the acidic solution and become positively charged. The positive charges repel one another and the protein molecules are less likely to combine with one another.

In a basic solution, the acidic parts of the protein lose a proton and become negatively charged. This also results in repulsion between the protein molecules, and combination is reduced.

Charged areas of the protein interact with water molecules because water is a polar molecule, with one end negative and one end positive. These charged ends are attracted to opposite charges on the protein.

Whether a protein forms a gel is thus affected by the acidity of the water it is dissolved in.

Milk

Nearly 80 percent of the proteins in milk are casein proteins. Caseins have a lot of the amino acid proline, which has a side chain that causes proteins to bend wherever there is a proline. This makes the proteins unlikely to stack into regular, orderly secondary structures. In addition, caseins have no disulfide bonds, and so they have little tertiary structure. This means that the hydrophobic parts of the molecule are open and exposed (not tucked inside a ball).

All of this combines to give caseins interesting properties. The hydrophobic parts end up migrating together, and the hydrophilic parts arrange themselves to the outside, facing the water. These tiny, hairy balls of protein are called *micelles*.

Caseins bind together with calcium and phosphorus. As a nutrient for the young mammal that needs to build bones, this is a useful property. Without the caseins, calcium phosphate would not be soluble. In the basic solution of milk, the hydrophilic parts of caseins become negatively charged and repel one another. This allows milk to stay liquid. But caseins clot in the stomach, due to acids that counteract the negative charges and enzymes that cut the proteins into smaller pieces. This clotting makes the proteins stay longer in the stomach, releasing amino acids slowly, which aids digestion and absorption of the protein.

In the stomach of young mammals, enzymes cut off part of the water-soluble casein (kappa-casein) that has the negative charges that keep the micelles apart. In making cheese, these enzymes, extracted from the stomachs of young calves, are used to make the caseins clot together into a solid.

Eggs

The proteins in eggs largely determine the characteristics of foods that contain eggs. Understanding the different properties of these proteins can be helpful when cooking or creating new dishes.

When you crack an egg into a pan, you can immediately see three parts. There is the yolk, a thin watery white, and a thick gelled white.

Egg white contains several mucoproteins, in which the protein is attached to carbohydrates. In the egg, these serve as nutrients for the growing embryo, and as support and protection.

Over half the protein in egg white is of one type: ovalbumin. It denatures at 176°F (80°C), forming the solid white mass you see at breakfast.

The next most prevalent protein in egg white is ovotransferin, also called conalbumin, which makes up about 12 percent of the proteins in the white of the egg. It denatures at a lower temperature, about 145°F (63°C).

The third most prevalent protein is ovomucin, at 11 percent. It is found close to the yolk, mixed with other proteins, thickening them.

When you crack an egg into a frying pan, the thin part of the egg white has less ovomucin, and the thick part of the white has two to four times as much. Ovomucin is the main gelling agent in egg white.

The first protein to denature when heating an egg white is the ovotransferin. As it unfolds, it binds not only to other unfolded ovotransferin molecules but to other proteins that are not yet denatured. The ovomucin molecules, which do not coagulate with heat by themselves, can thus be incorporated into a strong gel with the ovotransferin and ovalbumin.

The remaining proteins make up less than a quarter of the protein in egg white, but some of them bear mentioning here. Avidin makes up a very small portion of the egg white (less than a 1⁄10 of a percent), but it binds very tightly to the essential nutrient biotin (vitamin B_7), making the biotin unavailable as a food source. This effect is destroyed when the protein is denatured by heat or beating, but it can be a problem in a diet that contains a lot of raw egg white.

Meat

Raw meat is tough because each tiny packet of muscle fibers is surrounded by a tough sheet of connective tissue. This is the same tissue that, when boiled, makes gelatin.

When meat is cooked, the tough connective tissue denatures and becomes soft gelatin. The proteins in the muscle fibers also denature. Enzymes in the tissue no longer function when they are denatured, so cooked meat will keep longer than raw meat.

If the meat is overcooked, the water in the fiber bundles boils and the gelatin bag holding them bursts, and the meat dries out. At high temperatures, the proteins also undergo further denaturing and cross-linking, making the meat tough again. Crisp bacon is an excellent example of this. A thick, juicy steak would be inedible if cooked to the hardness of bacon.

Enzymes

Enzymes in foods are often a problem for food storage. As cells break open, the enzymes inside them leak out and react with other parts of the food. This causes brown soft spots in fruits and vegetables, and it makes meats smell and taste bad. The damaged parts also invite decay microorganisms.

Denaturing the enzymes can help to preserve the food. The heat of cooking is one familiar way to denature enzymes, but proteins can also be denatured by acids, strong alkalis, desiccation, or salt.

Shortening

When wheat flour and water are mixed and kneaded, sheets of gluten are formed. With further kneading, these sheets stick together into larger and larger sheets.

But if oils or fats are added, the hydrophobic amino acids in the gluten attach to the fat, so that they are not available to form bonds with other gluten molecules. This changes the nature of the dough, making it more tender and less able to trap bubbles of leavening gasses.

The result is a more cakelike, less breadlike structure.

Glutamate

One particular amino acid has a strong effect on the taste of foods. That is glutamic acid, and salts of it are called glutamates.

Besides being an abundant neurotransmitter in the brain, glutamate activates sensors on the tongue that detect savory protein-rich foods. Meats, poultry, fish, cheese, and soy sauce are rich sources of glutamate. The commercial form of pure glutamate is monosodium glutamate, MSG, which is an additive in many foods.

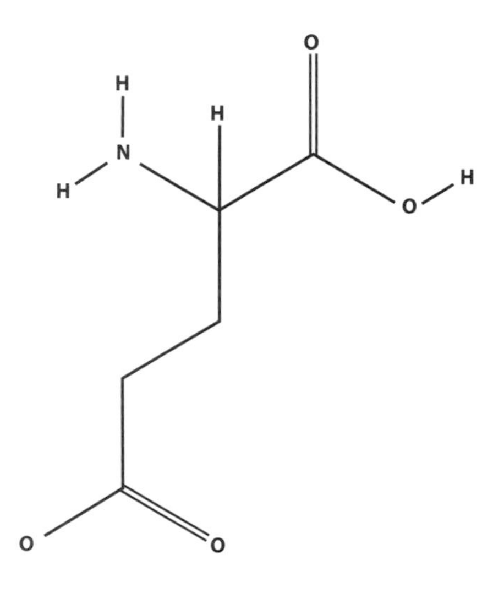

Glutamic acid

Cheese

Cheese is made from milk that has been curdled by the addition of acids and an enzyme from the stomach of calves called rennet. The acid can come from almost any food source, but for the most part it is produced by bacteria that convert the milk sugar lactose into lactic acid. Yogurt is also produced this way.

Cheese can be made without rennet, but the enzyme makes the curds stronger and more rubbery. Rennet allows the milk to curdle with less acid, which in turn allows flavor-producing bacteria to colonize the curd. Cheeses made with rennet will melt easily, while cheeses made with acid alone remain solid at high temperatures.

The curds are salted and moisture pressed out so the product will not be as easily attacked by bacteria as raw milk would. Thus making cheese is a way of preserving milk.

Recipe

Thanksgiving Turkey

Our feathered assistant, Corky, will help in the preparation, especially in the taste testing as the stuffing progresses.

Get up very early in the morning on the *day before* you plan to eat the turkey. It will be cooking all day and all night, and most of the rest of the next day. This bird will be cooking for an hour a pound, and this bird is 36 pounds. The turkey, not the parrot.

Because it will be cooking at a temperature below boiling, take some simple steps to ensure that nothing harmful grows in your bird.

The stuffing will be acidic and sweet, which will prevent bacterial growth. Dried fruits and nuts do not support bacterial growth. Other steps to prevent harmful bacteria will be shown later. These precautions are a good idea no matter how you cook your bird.

Ingredients:

- 1 pound butter
- 5 to 8 Pippin apples
- 2 cups dried nectarines (or apricots)
- 4 cups slivered almonds
- 4 cups sliced almonds
- 4 cups pecan halves
- 2 cups dried cherries
- 2 boxes commercial stuffing mix (about 12 ounces each)
- 2 eggs
- 1 quart apple cider (less for a smaller bird)
- 20- to 40-pound turkey
- 2 quarts hydrogen peroxide
- 4 pounds bacon

Supplies:

- 4-quart Pyrex bowl
- Huge bowl
- Large spoon
- Metal or bamboo skewers (or string)
- Casserole dish (for vegetarian stuffing)
- Paper bag
- Thermometer (or two)

Melt the butter in a 4-quart Pyrex bowl using the microwave.

Core and chop the apples into pieces about the size of a sugar cube. As you chop each apple, put the pieces into the melted butter. This will prevent them from getting brown on the cut edges.

Chop the nectarines into small pieces and add them to the apples.

Put the fruits and nuts into a huge bowl.

Add the two boxes of stuffing mix. I like to use one box of cornbread stuffing mix and one box of traditional. Mix everything well with a large spoon.

Add the two eggs, and mix very well, distributing the eggs throughout the stuffing. Add the apple cider.

At this point the stuffing is done. You and your assistant can now taste it. You might find you do a lot of tasting, as this is one of the best stuffings you will ever taste, and it needs no cooking to be enjoyed.

Now you are ready to wash and sterilize the turkey. Rinse it inside and out with hot water. Add a big handful of salt to the inside of the turkey, and use it as a scrubbing aid. Rinse out all of the salt, then use 2 quarts of hydrogen peroxide to sterilize the bird inside and out. Let the hydrogen peroxide sit inside the bird for a few minutes. Hydrogen peroxide kills any bacteria and molds that might be in or on the turkey. It breaks down into water and oxygen. You will see the oxygen bubbles scrubbing away inside the bird, and on the outside.

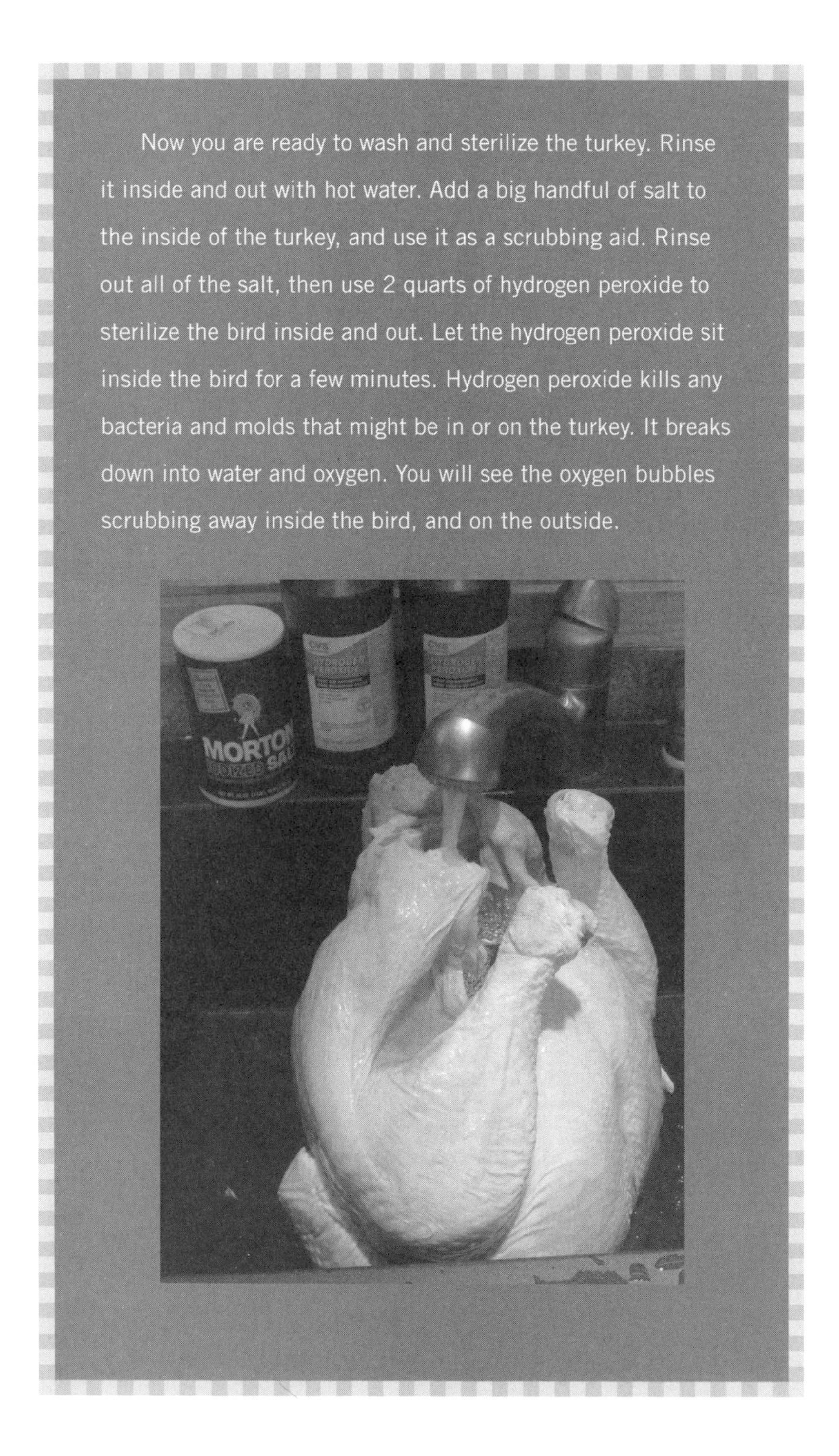

Don't rinse the peroxide off the bird. Almost all of it will drain out of the turkey as you tip it in the sink, but a little will remain inside as you stuff the bird. This will provide extra oxygen, which prevents the growth of botulism bacteria.

Now stuff both cavities of the turkey. Go ahead and pack it in tightly—this stuffing can take it.

Use metal or bamboo skewers to close up the openings. Some people lace them shut with string, but I find that the skewers work just fine by themselves.

Now make a basket weave of bacon on top of the turkey. I used to do a simple one-layer weave, but each year my guests keep demanding "More bacon!" and I accommodate them. You will see why it is so popular in a minute.

Depending on how large your turkey is, and how hungry you and your assistant are, there will be enough stuffing left over to put into its own casserole dish for your vegetarian friends. My assistant is more interested in the stuffing than the bacon-wrapped turkey.

Now take a clean paper shopping bag and cut out one side so you can place it over the turkey. This will prevent splattering in the oven and keep the steam close to the bird as it cooks.

I like to use two thermometers, one for the breast meat and one for the dark meat. It doesn't matter, however, as they both end up reading the same when the bird is done. I just like the comfort of redundancy.

A 36-pound turkey will just barely fit into a large oven.

The trick to a moist and tender turkey is to cook it at a temperature below the boiling point of water, but slightly above the temperature you want it to end up when it is done. I use a temperature of 205°F (96°C).

Cook the turkey at 350°F (176°C) for the first two hours to further sterilize the outside of the bird before you turn it down to 205°F (96°C).

In any meat, the tiny bundles of muscle cells are surrounded by a tough layer of connective tissue. To make the meat tender, raise the temperature until the connective tissue melts into gelatin. But if the temperature gets above the boiling point of water, the steam will burst open these little soft sacks of gelatin, and all of the juices will end up at the bottom of the pan instead of waiting to ooze out of the meat as it is sliced.

Because you have ensured that the bird is free of harmful bacteria, you can cook it to a lower temperature than is usually called for with poultry. This will prevent the breast meat from becoming dry. This is the way I like my turkey, cooked to about 150°F (66°C).

However, this results in the thigh meat looking red, especially around the joints, and this terrifies some of my guests. For them, I allow the turkey to cook to 160°F or even 170°F (71°C or 77°C), which makes the breast meat a little dry to my taste, but the dark meat is still moist and delicious. Being a dark meat fan, I find this compromise perfectly acceptable.

The turkey should be done a couple of hours before you start dinner. This allows you to use the oven for all of the other trimmings you want to have with the meal. More important, it allows time for the turkey to cool, and the little gelatin packets to gel, so the bird carves without falling apart.

For my guests, it also gives them time to pick off all of the bacon and eat it. This is not just an important ritual for this group, but it is also a necessary step before carving. All of that hard, brittle bacon shell has to be removed so the knife can find something to slice into.

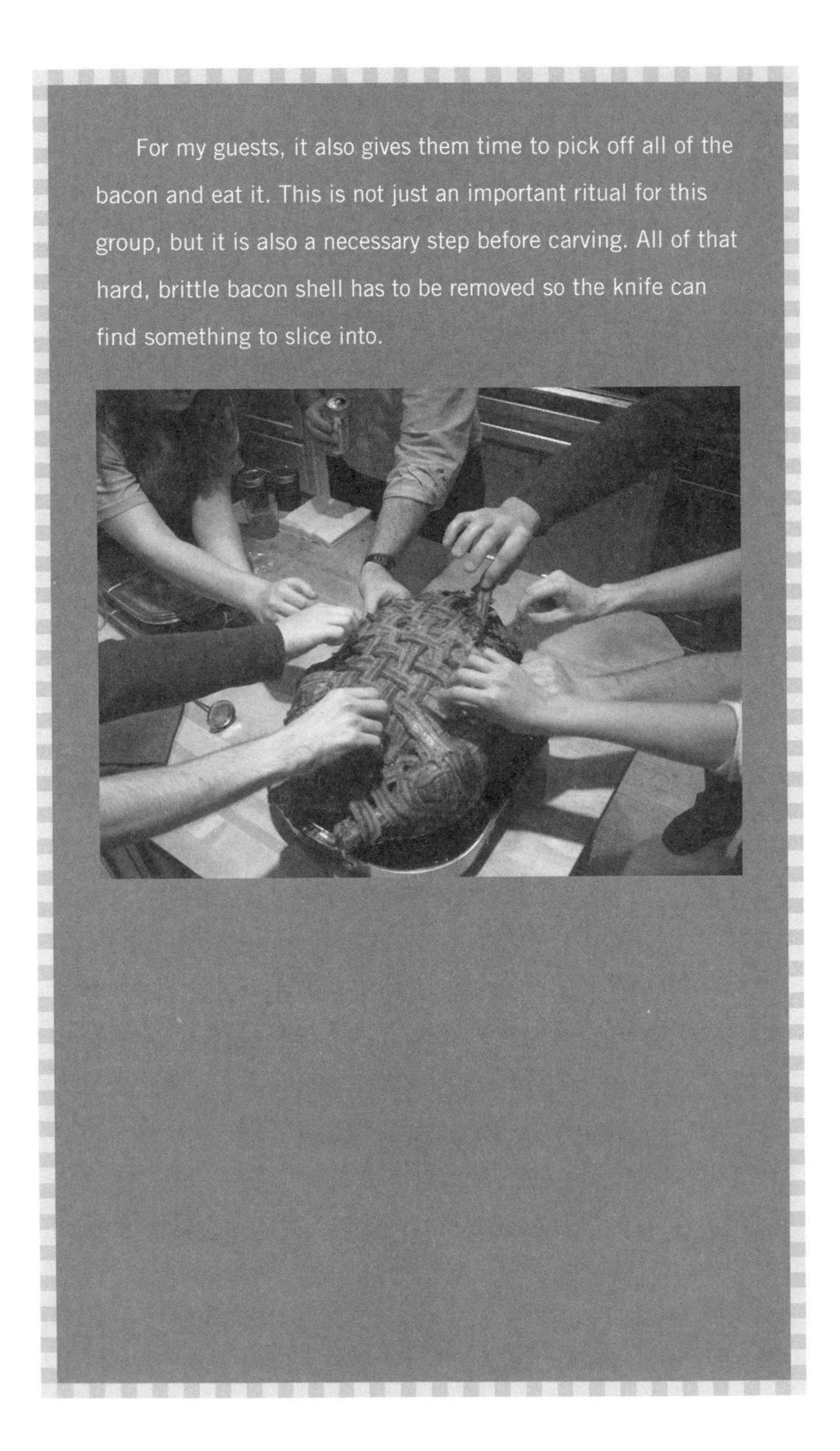

❖ 9 ❖

BIOLOGY

A lot of the food we eat is alive. Fresh, uncooked fruits and vegetables are made of cells that are still living and breathing long after they are picked. But cheese, yogurt, wine, beer, pickles, sauerkraut, olives, and many other foods are produced by the action of living microbes. In many of the foods we eat, these microbes are still alive, doing their job.

If you think for a moment about how difficult it is to keep something sterile, you will realize that just about everything has something living in it or on it. Most bacteria are benign, as are most yeasts, and there may be more beneficial varieties than there are harmful ones.

The yeasts and bacteria in a sourdough starter come from the flour used to make the starter (very little yeast is actually floating in the air). Bees carry yeasts from flower to flower, where they feed on the nectar. When the fruit sets, the yeasts grow in the skin of the fruit, where sugars and moisture are available. The powdery coating on grapes is yeast. Crushing the grapes allows the yeast to mix with the juices and ferment them into wine. No added yeast is actually necessary, but it is usually added to overwhelm any natural yeasts and produce a consistent product.

In most foods that are produced or modified by microorganisms, the environment is adjusted to suit the desired organisms and to make undesirable organisms have a harder time. In pickles, for example, salt is added to allow bacteria that produce lactic acid to outcompete harmful varieties that compete for the sugars in the cucumber. The bacteria then help out further by making the environment too acidic for other life-forms. Too much salt, however, allows harmful yeasts to grow, which then eat the lactic acid and thus invite even more undesirable microbes to grow.

Controlling other factors in the environment, such as temperature, moisture, and oxygen, also help select for the desirable organisms in the food you are producing.

Yeast

For thousands of years, people made bread with yeast, not realizing that they were farming microorganisms. That a living thing was responsible for the rising of bread was only recognized after the invention of the microscope.

Under a microscope, yeasts appear as single cells, like bacteria. Unlike bacteria, however, yeast reproduces by growing smaller yeast cells as buds on the parent cell. In some varieties of yeast, the buds do not detach, and long strings of cells form as the new buds grow to full size and then bud new yeast cells.

The yeasts used in bread, beer, and wine eat sugar. There is little sugar in flour, but uncooked flour has enzymes that break down the starch molecules in the flour into small sugar molecules that the yeast can feed on. The enzymes start to work as soon as water is added to the flour.

In wine, the sugars of the grape feed the yeast. In beer, some of the grain is allowed to sprout, and the sprouting seedling produces large amounts of the enzymes that convert starch into sugars. The sprouts are ground and added to the other grains in the vat, and the yeast feeds on the sugars that are produced by the action of the enzymes.

As the yeast feeds on the sugars, two waste products are excreted. One is alcohol, and the other is carbon dioxide. In beer, both are desirable. In wine, the carbon dioxide is often allowed to escape into the air before the wine is bottled. When a carbonated product is desired, extra sugar is often added just before corking the bottle so the yeast that remains can produce more of the gas.

In bread, it is the carbon dioxide that bakers are after. Alcohol is still produced, and a measurable amount of alcohol remains in the bread even after an hour of cooking at high temperature. This alcohol may help to dissolve aromatic molecules and enhance the scent and flavor of the bread, but generally you are only concerned with the amount of gas the yeast produces to make the dough into a nice light foam.

In modern bread recipes, commercially grown yeast is generally used. The large amount of yeast that is added to the dough ensures that this one strain of yeast outcompetes all other organisms in the dough, so that uncontrolled bacteria or wild yeast does not affect the flavor of the bread.

Yeast also produces enzymes such as transglutaminase that change the behavior of the gluten in the dough. The dough becomes less extensible as more of the enzyme is produced and more cross-linking occurs in the glutenin proteins. This strengthens the dough.

Yeast lives on small sugars such as glucose, sucrose, and maltose. The wheat kernel contains enzymes that break down the starch (its stored energy source) into maltose and glucose, which can then be used directly by the growing wheat sprout. Grinding the kernel into flour and wetting the flour allows the enzyme to break down the starch, so the yeast has a food supply.

The yeast has its own enzymes, among them maltase, which breaks down maltose into two glucose molecules. As the yeast feeds on the glucose, it produces not only carbon dioxide and alcohol but also many other molecules (aldehydes, ketones, aromatic heavy alcohols, and other metabolic byproducts) that add flavor to bread, beer, and wine.

Sourdough

Stir some water into flour and let it sit for a day. The microbes in the flour will begin to grow. Feed it more flour and water every day, and after a week the ecology in your jar will stabilize, with a mix of yeasts and bacteria that is called sourdough starter.

In practice, feeding a starter involves throwing half of it away when you add more flour and water. There is not very much food in flour that the microbes can actually eat, so throwing away the "used up" part of the mixture keeps it from diluting the new food with what the microbes consider waste matter.

After a week has gone by, and you have a nice stable culture, you no longer waste any starter. Instead, you can make bread with it and feed some of the resulting bread dough back to the starter jar.

Regular baker's yeast will not survive long in the acid environment of a sourdough starter. The variety of wild yeasts that come to live in harmony with the acid-making bacteria in the culture are of a different type, such as *Candida milleri*, *Candida krusei*, *Pichia saitoi*, and *Saccharomyces exiguus*. Along with these acid-tolerant yeasts grow acid-producing bacteria such as *Lactobacillus sanfranciscensis*, a species originally found in sourdough starter from San Francisco but later found to be in many sourdough starters all over the world.

The bacteria are smaller than the yeasts, and outnumber them 100 to 1 in most sourdough cultures. *Candida milleri* yeast can't eat the sugar maltose, but the bacteria can. In this way, they can coexist without competing with one another too much, even though both will eat glucose. The bacteria produce lactic and acetic acids and another antibiotic called cycloheximide that kill off many of the competing and potentially harmful organisms but leave the symbiotic yeasts and bacteria unharmed.

The bacteria also excrete glucose as they metabolize maltose, making the glucose available to the yeasts. Maltose is made of two glucose molecules bound together, and when the bacteria break them apart, some of the glucose escapes into the culture medium.

There are many different species and strains (varieties within a species) of bacteria besides *Lactobacillus sanfranciscensis*. Other lactobacillus species often found in starter cultures are *plantarum*, *pentosus*, *rossi*, *pontis*, *acidophilus*, *delbrueckii*, *homohiochii*, *hilgardii*, *viridescens*, *panis*, *pastorianus*, *oris*, *vaginalis*, *reuteri*, *buchneri*, *fructivorans*, *salivarius*, *brevis*, *fermentum*, *casei*, and *paraplantarum*. When present in bread, these bacteria act to inhibit mold and thus serve as preservatives.

The bacteria require more than just sugars to survive, and what they don't get from the flour they get from dead yeast cells. So each of the symbionts in the partnership produce something the other needs.

Of course there are many other yeasts and bacteria growing in the culture in smaller numbers. Some grow there despite the nonoptimal growing conditions of acidity and antibiotic substances produced by the main colonists in the culture.

A cook might want to control the growth of each of the two types of microorganism. The bacteria provide flavors that the cook might want to make more robust or more subtle, and the yeast provides the main source of carbon dioxide to make the bread lighter in texture. Knowing the best growing conditions for each type of microbe allows the cook to trade off one in favor of the other.

Candida milleri (the yeast) is tolerant of a wide range of acidity. The bacteria (*Lactobacillus sanfranciscensis*) grow best at a pH of 5.5, and don't grow as well below 4.5 or above 6.5. Adding vinegar or baking soda to the starter can select against the bacteria while allowing the yeast to grow (the cook would want to actually measure the pH to make sure it was in the desired range). The bacteria generally grow faster than the yeast at pH levels above 4.5.

The bacteria grow fastest at a temperature of about 90°F (32°C), while the yeast grows fastest at about 80°F (27°C). Since the yeast produces undesirable flavors when growing rapidly, a temperature above or below the optimum for speed of growth is considered best for bread flavor. So, if the cook wants to encourage the bacteria, a temperature of 90°F might be worth a try, and if the yeast is to be encouraged instead, a temperature of 70°F to 75°F (21°C to 24°C) might be sought. It might even be interesting to grow the starter at 90°F to get the flavors from the bacteria, and then "proof" the actual bread dough at 75°F to grow the yeast for faster rise times. The starter might be left to grow for a day unattended, but the waiting for the bread to rise might be less convenient.

The yeast is more tolerant of salt than the bacteria. The yeast will grow in salt up to a concentration of 8 percent, but the bacteria die out at 4 percent. Using that much salt in a bread recipe would not be advised, but since the starter will be diluted 5 or 6 to 1, this might be a place where experimentation can be done. Adding small amounts of salt has been shown to stimulate yeast growth in sourdoughs, possibly by eliminating some competition from the bacteria for nutrients they both require.

The yeast grows less as the alcohol content rises, but the bacteria are for the most part unaffected until the alcohol rises to 6 percent (the growth falls off sharply to zero in both organisms at 8 percent).

The yeast is strongly affected by acetic acid (which the bacteria produce from lactic acid in anaerobic conditions), while the bacteria are more tolerant. Both are fairly tolerant of lactic acid.

The reason these two organisms grow well in symbiosis is that they have very similar growth response at temperatures between 68°F and 82°F (20°C and 28°C) and pH levels around 4 or 5. When growing a sourdough starter (especially a new one), keeping the conditions in this range will select for a stable symbiotic mix and select against harmful organisms. Once the starter is well established, short-term deviations from these conditions to promote one type over the other can be successful.

When bacteria like *Enterococcus faecalis* and *Pediococcus pentosaceus* colonize a starter, some proteolytic (protein breaking) enzymes are produced. These harm the dough by breaking down the gluten needed to trap the gas bubbles. The lactobacillus bacteria do this as well, but to a much lesser extent. Nonetheless, selected strains of lactobacillus have been used to break down the gluten in wheat flours to make it palatable to people with celiac disease, a condition where the immune system is activated by gluten and attacks the lining of the intestine.

The breakup of the gluten protein is not completely harmful to the dough, however. Some of the breakdown products are important to the flavor and aroma of sourdough bread, and as the bacteria break down the protein to free the nitrogen they need for food, they convert some of it into essential amino acids normally missing from wheat flour, such as lysine. They thus improve the nutritional value of the bread to a small extent.

Yogurt

Adding *Lactobacillus acidophilus* bacteria to warm milk makes yogurt. The bacteria eat the sugar lactose in the milk and produce lactic acid. The lactic acid denatures the casein proteins in the milk, allowing them to unwind and bind together, forming a gel.

The bacteria compete with harmful bacteria that would otherwise cause the milk to spoil. The acidity of the yogurt prevents some of the harmful organisms from growing well enough to be a problem.

Raw milk can harbor a number of dangerous microbes, the most common of which are *Staphylococcus aureus*, *Campylobacter jejuni*, salmonella, *Escherichia coli*, *Listeria monocytogenes*, *Mycobacterium tuberculosis*, *Mycobacterium bovis*, brucella, *Coxiella burnetii*, and *Yersinia enterocolitica.*

Staphylococcus aureus grows best at neutral pH (7.0 to 7.5) but can survive in milk as acid as 4.5 pH. The pH of yogurt is just below that level (4.25 to 4.5). Likewise, salmonella, *Campylobacter jejuni*, *Escherichia coli*, *Listeria monocytogenes*, and *Mycobacterium tuberculosis* all grow poorly or not at all at pH levels more acidic than 4.5. Note, however, that these are not all of the dangerous organisms that can be found in raw milk, and pasteurization is a far better protective mechanism than culturing lactobacillus.

Sour Cream and Cultured Buttermilk

Lactobacillus is not the only bacterium used to sour dairy products. Both buttermilk and sour cream contain *Streptococcus diacetilactis*, *Streptococcus lactis*, *Streptococcus cremoris*, and *Leuconostoc citrovorum*, along with other less common bacteria.

Both buttermilk and sour cream keep for weeks in the refrigerator and are easy to make at home. If you have a container of milk that is getting old (but has not already turned sour), you can turn it into buttermilk for baking instead of waiting for it to sour and throwing it away.

The starter culture for buttermilk is some leftover buttermilk, since all of the necessary bacteria are already in the leftovers.

Make sure you have a very clean one-quart container. Put a cup of cultured buttermilk that has not passed its "use by" date (since you want the bacteria to be alive) into the container and fill it up with the milk. If making more or less, keep the 1-cup-to-3-cups ratio for best results.

Cover the container tightly and leave in a warm place for 24 hours. It should have thickened in that time, but if not, you can leave it for up to 36 hours before it no longer tastes good enough to drink (it is still good for baking, however).

The liquid left over from making butter is not the same as cultured buttermilk. It is not thick, and it often has flakes of butter in it. In the United States it is referred to as "old-fashioned buttermilk." Some cultured buttermilk has bits of butter added to mimic old-fashioned buttermilk. But only true cultured buttermilk will work well as a starter for the buttermilk you need in baking, or for buttermilk pancakes.

Making your own sour cream is as easy as making your own buttermilk. Add a small amount of buttermilk to some cream and let it sit in a warm place for a day. The higher the fat content of the cream, the thicker the sour cream will be.

Bleu Cheese

When Alexander Fleming discovered penicillin on September 3, 1928, it may have been more than just his untidy laboratory that deserved the credit. It may have been his lunch.

He had been studying staphylococci, a disease-causing bacteria, and had stacked his cultures in a corner of his lab when he went on vacation for the month of August. On returning September 3, he noticed that one culture was contaminated with a fungus and that the bacteria around the fungus had been killed.

It took another 10 years before chemists at Oxford were able to produce a stable form of the antibiotic from the fungus, and 5 more years before industrial methods were developed to produce it in usable quantities.

But penicillium molds had been used as antibiotics since ancient times, long before bacteria were known to be the cause of many diseases. In 1870, it was noted that cultures with mold in them would not produce bacteria. John Tyndall demonstrated to the Royal Society in 1875 that penicillium fungus had antibacterial properties. In 1877 Louis Pasteur observed that anthrax was inhibited in cultures that had been contaminated by molds. In 1897 Ernest Duchesne (using a different species of penicillium than that used to make penicillin) cured guinea pigs of typhoid.

Fleming isolated *Penicillium notatum* as the species that produced the antibiotic, and later work identified strains of that species that produced even more of the substance. Duchesne used *Penicillium glaucum*, the species used to make Gorgonzola cheese. That species, along with *Penicillium roqueforti*, is also used to make bleu, Roquefort, and Stilton cheeses.

Penicillium is a common bread mold, and it is the main cause of spoilage in stored grains. Like many of the other organisms we have been discussing, it produces chemicals that inhibit or kill competing organisms. Unlike the alcohol or lactic acids produced by yeasts and bacteria in wine and yogurt, the chemicals produced by the mold are not a waste product that has side benefits, but has evolved because it gives the mold a strong advantage in competing for food.

In cheese, it not only provides the characteristic flavor and aroma of bleu cheese, but it also protects the cheese from spoilage from more harmful bacteria and fungi that would either produce undesirable flavors and odors or produce toxins that would make the food unfit for consumption.

Alexander Fleming was studying staphylococcus cultures to find something that would kill them, so his rediscovery of the benefits of the mold came when he was well prepared to make use of the information. But it may have been the bread or cheese from his lunch, left in the lab for a month while he was on vacation, that should actually get the credit. We'll never know.

Roquefort cheese contains the highest levels of glutamates of any naturally produced food. Glutamates are what give savory, protein-rich foods their taste, and they are found in other fermented foods, such as soy sauce. Purified as monosodium glutamate and used as a flavor enhancer, it is sometimes overused, and people with a sensitivity to it complain. But it is found in almost all foods that contain protein, and our tongues have specialized sensors for it to help us find nutritious, protein-rich additions to our diet.

You can make your own cheese fairly easily these days, thanks to the availability of rennet tablets in the supermarket. Adding a little bit of bleu cheese or Roquefort to the curds before pressing the cheese can be a tasty way to experiment with microbial processing and the use of antibiotics to preserve foods.

Wine and Beer

Yeast is on the outside of grapes and grains such as barley, wheat, and rice. So it is no surprise that grape juice or wet grains eventually ferment. Beer and wine have been around since before agriculture, for more reasons than just to provide a head buzz.

As with other foods made with microorganisms, creating a friendly environment for one organism can make that same environment less friendly to other, potentially harmful organisms. The alcohol that yeast produces helps to preserve the wine and beer, so those calories and nutrients can be saved for a day when food is less available.

As people began to live in more concentrated populations, the lack of sanitary drinking water became a problem. Septic systems and sewage treatment plants were not available, and the local water supply was not always fit to drink. But wine and beer, even with low alcohol content, were generally safe to drink. Either the water came from the grape or it was boiled with the grains. In either case the germs that cause typhus, diphtheria, dysentery, and other water-borne diseases were not present.

There are between 700 and 900 calories in a liter of wine, and 360 or so in a liter of beer. So besides being safer to drink than the local water, they provided energy and some amount of nutrition.

Beer has been called "liquid bread" because it is made of grains and yeast. Weak beers with low alcohol content could be consumed without compromising mental faculties or losing bodily fluids from the diuretic effects of alcohol.

Making beer and wine is simple. Making *good* beer or wine is an art. Like most of the foods made with microorganisms, much of the work is involved with making sure you encourage the organisms you want and discourage those you don't want.

Sterilization helps, so that there are no harmful organisms to start out. Controlling the temperature also helps to make a consistent brew. Paying attention to the nutrient requirements of the desired organism, and limiting the nutrition of the undesired ones, will help to keep the ecology of the brew in balance. Controlling the mineral content of the water used (in the case of beer) affects both the taste of the resulting beer and the health and vigor of the yeasts that make it. Making sure the brew gets enough oxygen helps to keep the yeast healthy and to prevent harmful anaerobic bacteria from growing. Brewers carefully control when aeration occurs. The yeast multiplies when there is enough oxygen, and it produces alcohol and carbon dioxide when there is less. Aeration when the temperature is too high causes bad flavors to develop.

Controlling the temperature helps to promote yeast growth and inhibit bacterial growth. The *wort*—the liquid that becomes beer after fermentation—is boiled but then quickly cooled, since undesired bacteria grow well at temperatures between 130°F and 90°F (54°C and 32°C). Aeration is begun when the temperature is below 80°F (26°C) to prevent oxidation while still providing oxygen for the yeast to multiply. Rapid cooling also prevents sulfur-containing compounds from redissolving in the wort, causing a cooked cabbage taste.

In beer making, there are two main types of yeast: top fermenting and bottom fermenting. Top-fermenting yeasts, called *ale yeasts*, prefer warmer temperatures; they go dormant below about 55°F (12°C). Bottom-fermenting yeasts, called *lager yeasts*, stay active down to about 40°F (4°C). While lagers are usually brewed at lower temperatures than ales, steam beer uses lager yeasts at the higher temperatures to get a different flavor.

To keep foreign organisms out of the beer as it ferments, the gas is allowed to escape through a tube to bubble up through water. This is called an air lock.

As with other foods produced by microorganisms, it helps to start with a large number of the desired organisms to overwhelm any of the undesirable ones. In the case of beer and wine, yeast is often added rather than relying on the natural yeast on the grapes or on the grain.

Brewers use a device called a hydrometer to measure the density of the liquid. This indicates how much sugar is dissolved in the wort, and thus how the fermentation is progressing (as the yeast consumes the sugar and produces less dense alcohol).

To make sparkling wine or control the carbonation in beer, sugar is sometimes added just before bottling. This allows the small amount of remaining yeast to produce carbon dioxide, which cannot escape the bottle. The pressure rises and the gas dissolves in the water.

Preserving

Many of the processed foods we enjoy were invented as ways of preserving fresh foods for consumption later. Dried fruits, cheese, salted meats, fermented beverages, yogurt, pickles—all are ways of extending the useful storage time of fresh fruits, grains, meats, and dairy products.

Salt and Drying

Life needs water to survive and grow. Eliminating water prevents harmful organisms from consuming the food and spoiling it. Many fresh fruits and meats can be simply dried in the sun to preserve them. Fresh-cut grass is dried to store as livestock feed.

Salting food is another way of preventing harmful organisms from getting the water they need to survive and thrive. If the outside environment is saltier than the inside of an organism, the water diffuses out to dilute the salt. The organism dries out, even if there is lots of salty water outside.

Meat and fish are salted more than fruits because the sugar in the fruit has the same osmotic effect on harmful organisms as salt does, once the fruit has dried enough to concentrate the sugar a bit. Fish and meats spoil quickly in the sun without some salt to prevent bacteria, molds, and yeasts from getting a foothold before the drying has progressed enough to preserve the food.

Heat Sterilization and Smoking

Another way to prevent organisms from getting started is to keep the temperature high enough to kill them or prevent growth. This can be done while drying the food, or in the case of canning, it can be done while the container is sealed to prevent organisms from getting into the packaging after the food has been sterilized.

Smoking preserves food by heating, drying, and sealing the outside of the food with substances that do not promote the growth of spoilage organisms. Smoking foods also controls oxidation to some extent, which is another way that food becomes spoiled. The smoke itself is only a coating; it does not penetrate very far into the food.

Alcohol Sterilization

You have seen how alcohol preserves grains and fruit juices when they are fermented, but alcohol is also sometimes used to preserve other foods. In brandied fruit, alcohol is added to chopped fruits that are then allowed to ferment. The alcohol prevents bacteria from competing with the fermentation yeasts in the early stages of the process. Once the process has been started, new fruit can be added as the brandied fruit is eaten, without needing to add extra alcohol. In this way, the original brandied fruit acts like the starter used in making sourdough bread.

Antimicrobials in Herbs and Spices

It is no accident that the use of spices in food correlates with the average temperature of the region from which the recipe is derived. Temperatures that promote the growth of bacterial and fungal spoilage organisms are prevalent in the countries where foods are heavily spiced.

The essential oils in sage, mint, hyssop, and chamomile have *bacteriostatic* effects—they prevent bacteria from growing—on Gram-negative bacteria such as E. coli and Gram-positive bacteria such as *Listeria innocua*. The essential oils from oregano have bactericidal (the bacteria are actually killed) effects that are most pronounced on Gram-negative bacteria.

Strong antimicrobial activity is found in cinnamon, cloves, and mustard. Less pronounced effects in allspice, bay leaf, caraway, coriander, cumin, thyme, and rosemary are similar to those of sage and oregano. Black pepper, red pepper, and ginger have only weak activity.

Garlic and onion are effective against salmonella, E. coli, staphylococcus, and *Candida albicans*.

Cinnamon, cloves, and sage all contain eugenol, an effective antimicrobial and mold inhibitor that is used in mouthwashes and baked goods.

Sage and oregano contain thymol, another effective antimicrobial, often used in toothpastes.

The oleoresins in rosemary also act as antioxidants and prevent rancidity of fats and oils in foods.

Acids

You have seen how making foods more acidic helps to preserve them, as in the case of yogurt and sourdough bread. But raising the acidity much higher—by pickling foods in vinegar—is an ancient and effective way of preserving foods from olives and cucumbers to boiled eggs and herring.

Besides yogurt and sourdough bread, other foods are pickled by fermentation that produces lactic acid. The most recognizable is probably pickled cucumbers or peppers. But sauerkraut and kimchi are also fermented in this way, after adding salt to make a brine.

Microbial Competition

Friendly microbes are also used to help fight harmful ones. Some produce toxic chemicals such as alcohol or lactic acid, but others produce more targeted antibacterial agents. The penicillin molds in some cheeses kill bacteria. Bacteria in other cheeses produce propionates that kill molds and fungi.

As you have seen before in several foods, controlling the environment to favor the organisms we like helps to limit the growth of the organisms we wish to avoid.

Recipe

DNA from Your Halloween Pumpkin

Toward the end of October you'll start seeing lots of recipes for pumpkin this and pumpkin that. This one is different. It involves liquor and a blender, but it is something I recommend you do with kids. It takes very little time to make, but your kid might want to take it to school for her science fair project. That said, you probably won't want to eat it. But it's fun.

You're going to extract the DNA—the very stuff of life—from a pumpkin.

Ingredients:

- Some alcohol, the closer to 100 percent alcohol, the better (see the instructions for more advice on your alcohol options)
- 1 pumpkin (this recipe will also work with apples, bananas, squash, carrots, or just about anything else you have in the fridge)
- 1 tablespoon salt
- ¼ cup dishwashing detergent (preferably the kind that has nothing in it but the detergent to ensure that weird additives won't spoil the experiment, but the stuff you have in the kitchen will probably work fine even if it is dark green and smells funny)

Supplies:

- A coffee filter (I use a Melitta #2 filter, but a paper towel folded into quarters will also work fine)
- A funnel (you can use a coffee maker in a pinch, but rinse it out when you're done)
- A shot glass or spice jar

For the high-percentage alcohol, you can use rubbing alcohol, which is about 70 percent, or you can get 91 percent isopropanol at some drugstores—that's what I normally use. But you might already have some 151-proof rum or some Everclear grain alcohol close to the kitchen. Not being a drinker, I had to go out and buy some for the photos. (Someday I'll cook up something that needs to be set on fire to use up the remainder of the bottle.) Put the alcohol in your *freezer* right away. You will want it as cold as you can get it for the experiment. Don't worry, it won't freeze—you can keep it in the freezer overnight or forever. But an hour in the freezer will probably do fine.

Cut a pumpkin into small chunks that fit in the blender. My assistant, Corky, helps me with all my Halloween projects. He also likes to eat raw pumpkin, something your helper might not find so tasty. Have your assistant place the pumpkin chunks in the blender, as shown in the photo.

Add about ½ cup of water to the pumpkin chunks, and 1 tablespoon of salt. Blend until most of the chunks are pureed. You don't need to be perfect, or blend forever, since you are going to filter later, and blending too long will break the DNA strands. But you want something resembling apple-sauce when you're done. Your assistant is allowed to taste the results.

Now add a whole bunch of dishwashing detergent, about ¼ cup. Your assistant is no longer allowed to sample the recipe after this point, but may look at it kind of funny. The detergent and the salt will break open the cell walls of the pumpkin, releasing the DNA into the water.

Next, filter the puree using the coffee filter and funnel.

Now it is time to get the liquor out of the freezer. You did remember to put it in there at least an hour ago, right? Your assistant is not allowed to sample the liquids, but may continue eating the raw pumpkin.

For this part, a shot glass would be appropriate. Too bad I don't have one; I used a spice jar instead. You could have just used the plastic cup, but a smaller diameter container, such as a test tube, is normally used. Pour some of the filtered pumpkin juice into the container.

Now, tilt the container and *very gently* pour some of the alcohol slowly so that it forms a layer on top of the juice but does not mix.

After a few seconds, you will see a ghostly layer of DNA form between the juice and the alcohol. It looks like ghost ectoplasm—a white slimy band of goo floating in the middle of the shot glass or spice jar. This little layer of molecules is the stuff that tells the cells to make a pumpkin and not a banana or a monkey. The pumpkin inherited it from its parent pumpkin plants.

When scientists extract DNA from other plants or animals to study it, they use techniques very similar to what you just did.

10

SCALING RECIPES UP AND DOWN

For most recipes, scaling up or down is just arithmetic. If a recipe serves four and you have six dinner guests, multiply each ingredient by 6/4 and you will generally get good results.

However, not all recipes are this easy, and scaling a recipe for four people up to a banquet of 200 people by multiplying by 50 probably won't work out so well. Why that is the case gets us into some interesting science. And the science can get us safely back to scaling the recipe.

Surface-to-Volume Ratios

One thing that causes problems in recipe scaling is that as you increase the volume of the ingredients, the surface area of the mixture does not increase as quickly. As an example, suppose you have a recipe that calls for boiling a quart of chicken stock in a four-quart saucepan. When you double the recipe, the bottom surface of the saucepan, where the heat is, remains the same. The top surface, where the evaporation happens, remains the same. The depth of the liquid is the only thing that has doubled along with the volume.

If your saucepan is eight inches in diameter, a quart of liquid will fill it a little over an inch (call it 1.14889975 inches if you like). Two quarts raises the level to almost 2⅓ inches (2.29779949 inches). You can do all the arithmetic easily using Google: just search for "2 quarts / (pi * 4 inches * 4 inches) in inches."

Calculating the surface area can be done the same way:

- One quart: Google for "(2 * pi * 4 inches * 4 inches) + (2 * pi * 4 inches * 1.15 inches) in square inches."
- Two quarts: Google for "(2 * pi * 4 inches * 4 inches) + (2 * pi * 4 inches * 2.3 inches) in square inches."
- Ratio of increase: Google for "([2 * pi * 4 inches * 4 inches) + (2 * pi * 4 inches * 2.3 inches]) / ([2 * pi * 4 inches * 4 inches) + (2 * pi * 4 inches * 1.15 inches])."

For one quart, you have 129 square inches of surface area. For two quarts, you have 158 square inches. The volume has doubled, but the surface area has increased by only 158/129, or about 1¼ times.

In a saucepan, this will mean several things. Since the heat applied to the bottom is the same, and the evaporation from the top is the same, it should take about twice as long to bring it to a boil, and about twice as long to reduce the stock to half its volume.

This is not a big problem for reducing a pot of chicken stock, but consider what happens if you are baking a loaf of bread and you double the volume. If you make the loaf twice as long, you don't have much of a problem. Likewise, you can make two loaves and not have much trouble. But if you try to keep the shape of the loaf by scaling up the length, width, and height by the same amount, you get into trouble.

Suppose you have a loaf that is 5 inches wide, 4 inches high, and about 8½ inches long. It has a volume of 3 quarts. Doubling each dimension to 10 inches wide, 8 inches high, and 17 inches long gives us a volume of almost 24 quarts. That's eight times the volume.

Should you cook it twice as long, or eight times as long?

Heat Flow Rates

In an oven, the heat comes from all sides of a loaf pan. The heat has to travel twice as far to reach the center of a large loaf. But there is eight times as much loaf to heat, and you have only twice the surface area through which the heat can reach the dough.

Being a foam, bread dough is not a particularly good conductor of heat. The outside of the loaf will dry out and then brown before the inside has reached the temperature needed to make the starches gel and the proteins denature.

The rate at which something heats up is proportional to the difference between the temperature inside and the temperature outside. This is known as Isaac Newton's Law of Cooling, although it works the same for both heating and cooling. As a loaf of bread heats up, the difference between the inside and outside temperatures gets smaller, and so the rate at which it warms gets slower.

Small things heat up quickly. Large things heat up more slowly. So as a recipe is doubled, the rate at which the bread cooks goes *down*. But because bread is not a good conductor of heat, the outside cooks faster than the center. This gets worse the larger the loaf. You can reduce this effect by cooking at a lower temperature, taking advantage of Newton's Law of Cooling.

Solving the Surface-to-Volume Problem

What makes the volume grow faster than the surface area is the number of dimensions involved. As the radius of a sphere doubles, the surface area goes up four times, because while the radius is a one-dimensional line, the surface area is two dimensional. You multiply the radius by two, but you multiply the surface by two because the width doubled, and then by two again because the length doubled. The volume is three dimensional, so you multiply by two a third time (that's a factor of 8 because $2 \times 2 \times 2 = 8$) to account for the depth doubling. This is why surfaces are measured in square inches, and volumes in cubic inches.

If we want to keep the surface-to-volume ratio constant, you can eliminate that third doubling by keeping the depth constant. This almost works perfectly. A 2-inch cube has a surface area of 24 square inches, and a volume of 8 cubic inches, for a ratio of $3/1$. Doubling the width and length gives a volume of 32 cubic inches and a surface area of 64 square inches, for a ratio of $2/1$.

Not perfect, but much better than doubling all the dimensions, where the ratio would go from 3/1 down to 2/3.

To make up for the rest of the difference, you can reduce the temperature and cook for a longer time.

Consider baking a wedding cake, for which you want to use the same batter for all the layers, but you want the top layer to be 6 inches in diameter and the bottom layer to be 16 inches in diameter, with other layers in between of successively increasing diameter from top to bottom. Each layer will be 4 inches in height, composed of two layers, each 2 inches high, cemented together with frosting or jelly.

You do some testing and find that a 6-inch cake is best when baked at 350°F (177°C) for 30 minutes, and a 16-inch cake is best when baked at 325°F (163°C) for an hour. Instead of testing each intermediate layer, you remember some algebra for finding the formula of a line given two points, and use that to draw a graph.

Now you can just look up the diameter of the layer on the bottom axis and read off the temperature and cooking time according to their lines. And, as a bonus, you can find out how many cups of batter each layer will require. As the diameter of the layers grows, the temperature gradually decreases, and the time needed to bake it gradually increases.

Drying

The surface-to-volume ratio affects how long it takes something to dry. Water evaporates at the surface, so the surface area determines the rate at which something dries. But the amount of water in something depends on the volume.

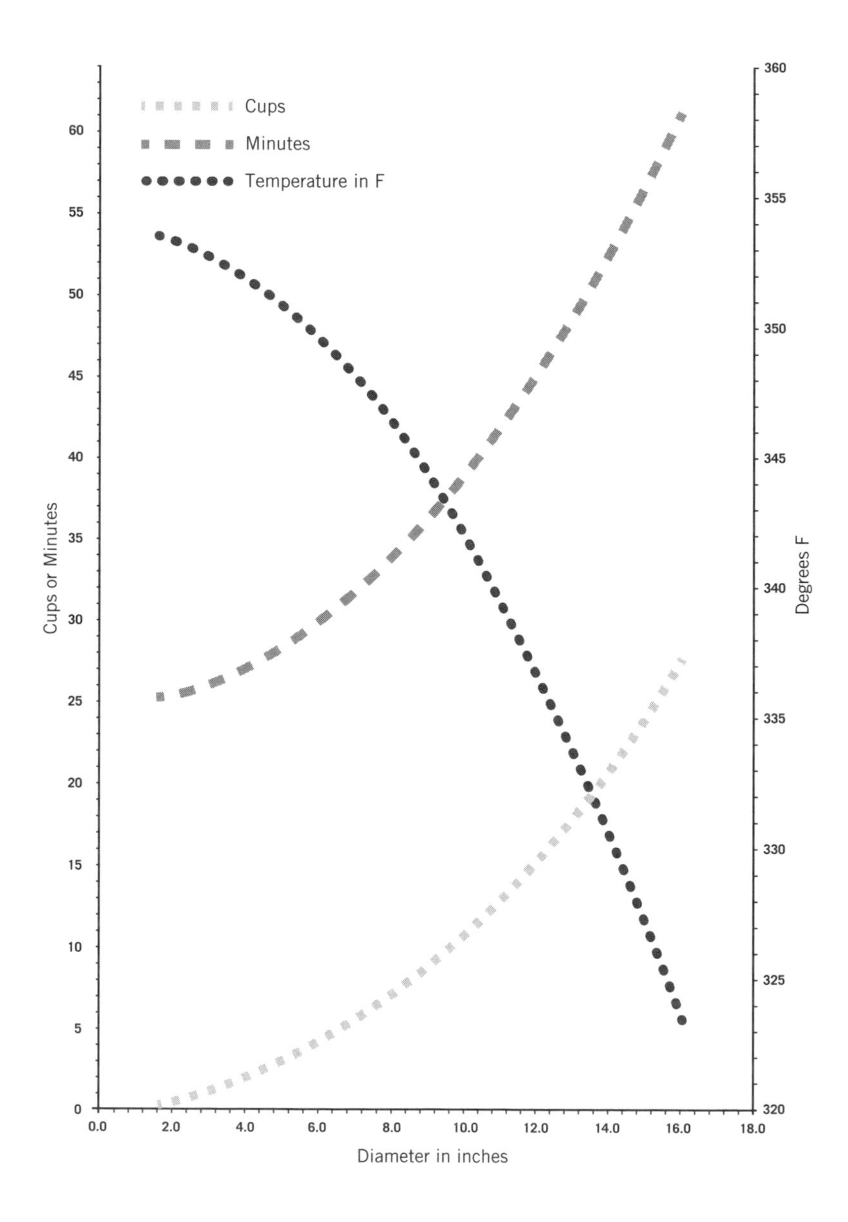

Wedding Cake Variables
Cups
Minutes
Temperature in F
Cups or Minutes
Degrees F
Diameter in inches
0
5
10
15
20
25
30
35
40
45
50
55
60
320
325
330
335
340
345
350
355
360
0.0
2.0
4.0
6.0
8.0
10.0
12.0
14.0
16.0
18.0

If you want something to dry quickly, spread it out over a large area. If you want something to stay moist, form it into a sphere (the shape with the smallest surface-to-volume ratio), or put it into a tall cylinder (assuming you can't just cover it up!).

Sea salt is evaporated in very shallow ponds that have square miles of surface area. Sun-dried apricots and prunes are spread out to dry, as are cacao beans and coffee beans.

The smaller an object is, the higher the surface-to-volume ratio. So finely ground bread crumbs will dry faster than a slice of bread, and a thin slice of bread will dry faster than a thick slice.

In making dried apples, slicing the apple into very thin slices and drying them in a 250°F (121°C) oven for an hour will make crisp chips. Trying to dry the whole apple will be much less rewarding, and take much longer.

Timing

You saw how adjusting the timing when baking different sizes of cake allowed you to scale up the recipe, and how you can speed up drying by increasing the surface-to-volume ratio. Other things are also affected by or affect timing.

Deep-frying a large potato will not get the same result as deep-frying thin slices of potato. But deep-frying a big bucket of french fries will not produce the same result as dividing the bucket into quarters and frying four batches. Dumping a large amount of cold food into hot oil cools the oil too quickly. Newton's Law of Cooling works both ways. It has the same effect in baking. Putting several loaves in the oven at once cools the oven, and each loaf shades the other from one of the hot oven walls.

You can compensate in several ways. You can raise the temperature of the oven. You can blow the hot air around inside the oven, as in a convection oven, so that the food cooks more from the hot air than from the heat radiated from the oven walls. You can cook for a longer time. You can also put big slabs of stone in the oven, called *thermal mass*, that preheat along with the oven walls and take longer to cool. The extra thermal mass counters the mass of the cold food.

Gravity

Doubling a recipe that involves foam creates another problem besides the low thermal conductivity of the foam: the weight of the top half of the food presses down on the bottom half, and the bubbles get progressively smaller the closer you get to the bottom. This makes the bottom denser than the top, so they cook at different rates.

Equipment

You may not always have a pot or pan that is exactly the right size. For a soup, this may not matter much, as long as the bottom of the pot is evenly heated. Making a soup in a skillet is possible, but it may require more stirring to prevent the food from sticking to an unevenly heated bottom, and it will lose water faster, since there is more surface for evaporation.

If the stove is capable of delivering heat effectively to the larger surface, the soup may actually cook faster. This may not be desirable if the flavors need time to blend, or it may be just the thing for reheating yesterday's stew.

In baking, you can use what you've learned earlier to adjust a recipe to the size of pan you have available. Pay attention to the surface-to-volume ratio, and adjust the temperature and the cooking time if you can't keep the ratio the same as the recipe calls for. If your pan is too small, consider making the dish in two batches. If it is too large, you might scale up the recipe and freeze what you don't eat today.

Different pots and pans conduct heat differently. A cast iron skillet does not conduct heat as well as copper or aluminum, and it may develop hot spots where the heat is applied. You can test a frying pan for heat conductivity and evenness of heating by sprinkling a fine layer of flour in the pan and heating it. If the flour browns evenly all over the bottom of the pan, the combination of stove and pan are effective at spreading the heat. If some areas brown first, you can see where the hot spots are. That pan will need more careful stirring when used for anything but boiling water.

The cast iron pan is much heavier than an aluminum pan or a thin stainless steel pan with a copper bottom. Once it gets hot, it will retain the heat longer due to this extra thermal mass. If you want to present your fajitas at the table still sizzling in the pan, the cast iron pan is the one you want.

11

We heat food for many reasons; cooking the food is only one of them.

Water for tea or coffee is heated because the volatile oils and flavors dissolve better in hot water, or are released into the water when fats melt or membranes break. Volatile components that are vaporized by the heat then reach our nose, creating aroma and flavor.

Heating starches changes crystallized starch molecules into gels. Bread becomes stale when the starches crystallize, and warming the bread returns them to their soft gel state, making the bread taste and feel fresh. Stale bread is not dry; it just feels that way because of the crystallized starches. In raw potatoes, the starch is compact, but heating makes the starch granules swell and absorb water, becoming soft and easier to digest.

Heating meat causes the tough collagen connective tissue to denature and soften into a gel. Heating it more causes the other proteins to harden, and we get crisp bacon.

Browning Reactions

There are two ways that foods become brown: through the action of enzymes, as when a sliced apple turns brown, and without enzymes. In the latter category, you can break the browning process down further into three groups: caramelization, where sugars react with one another; ascorbic acid oxidation, where vitamin C reacts with oxygen; and Maillard reactions, where sugars react with amino acids.

The Maillard reaction is complex. The reaction of the sugar glucose and the simple amino acid glycine gives more than 24 reaction products. In many foods, there are five or more sugars and as many as 20 or more amino acids, and they can all react when heated to create the brown color found on toast. That color is made up largely of *melanoidins*—large molecules that polymerize from the products of the Maillard reaction. Melanoidins have antioxidant properties, like many other food colors, and can bind to metal ions like iron and take them out of solution (a process called *chelation*, named for the chelae, the claw of a crab, since they grab onto the ions).

Five-carbon sugars such as ribose react more easily than six-carbon sugars like glucose, or disaccharides such as sucrose or lactose. The amino acid lysine produces the most color when reacting with sugars, and cysteine produces the least color. Foods that are rich in lysine, such as milk proteins, brown easily. This is why milk is used as invisible ink. It does not take much heat to make the milk turn brown, so the ink browns before the paper does.

Maillard reactions produce flavors as well as colors. Melanoidins have their own flavors, but their long chains full of sticky-side groups allow them to hold onto smaller flavor molecules that arc also produccd by thc Maillard rcaction, such as isobutyraldehyde and furfural, and hydroxymethylfurfural. These are then slowly released into the air as the aroma of toast, coffee, beer, and other foods that owe much of their flavor to the roasting or steeping process.

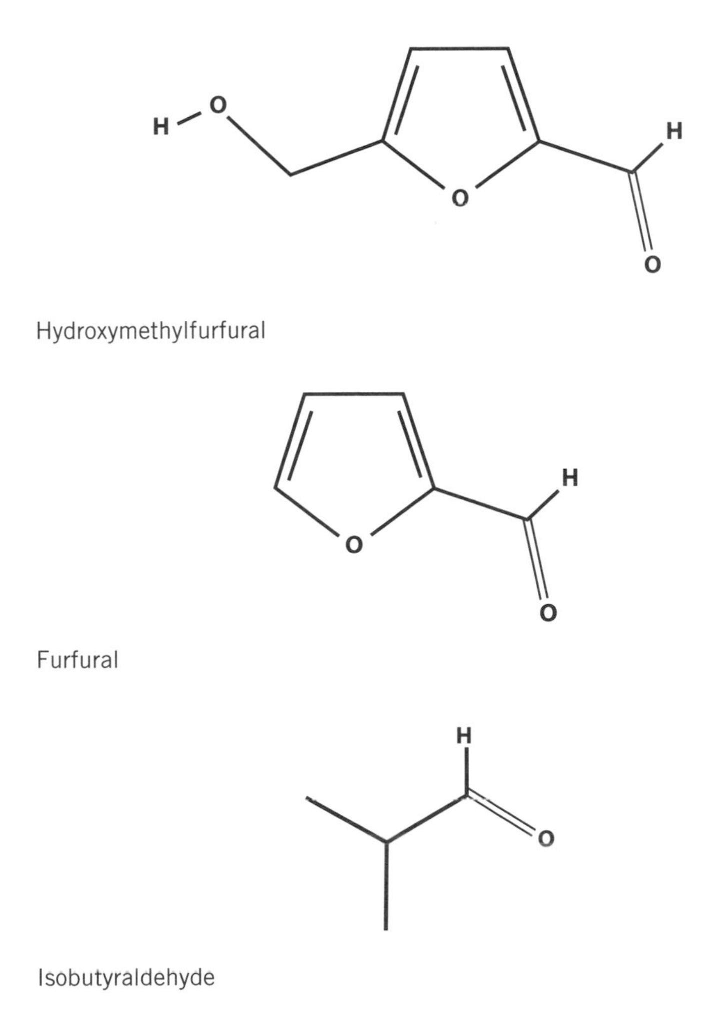

Hydroxymethylfurfural

Furfural

Isobutyraldehyde

Maillard reactions happen at room temperature (one of the reasons soils are brown), but they happen faster as temperatures rise. Above 248°F (120°C), the sugars in food combine with one another in the process known as *caramelization.* Fructose caramelizes at the lowest temperature, about 230°F (110°C). Maltose doesn't caramelize until it reaches about 356°F (180°C).

Caramelization is another complex reaction, involving the breakdown of complex sugars into simple sugars, followed by polymerizations into larger molecules, oxidations, isomerizations (changing the shape of the molecule without changing the number and type of the atoms in it), and other reactions. The result is a brown color and the familiar odors and flavors of burned sugar.

As sugar is cooked, it first breaks down into glucose and fructose, and then these simple sugars combine by losing water molecules (an OH group on one molecule reacts with an H on the other, and they join up as the resulting H_2O molecule boils away). The result is a type of molecule called a sucrose *anhydride* (meaning "without water").

Heating sucrose at 392°F (200°C) for 35 minutes results in the loss of one water molecule for every molecule of sucrose, producing a molecule called *isosacchrosan.* After another 55 minutes, the total water loss is four molecules per molecule of sucrose, and the molecule *caramelan* is formed. Caramelan is a bitter-tasting molecule with the formula $C_{24}H_{36}O_{18}$. Yet another 55 minutes of cooking causes every three sucrose molecules on average to lose eight molecules of water, and the molecule *caramelen* is formed, with the formula $C_{36}H_{50}O_{25}$.

Heating the sugar further creates an even bigger molecule that is very dark and does not dissolve well, called *caramelin*, with the formula $C_{125}H_{188}O_{80}$.

Protein Denaturing

Heating proteins destroys their carefully folded, three-dimensional structure and makes them no longer able to perform their normal functions. Egg white becomes a firm opaque solid, collagen loses its ability to connect bone and muscle and becomes gelatin, and enzymes lose their ability to efficiently catalyze chemical reactions. In the process, eggs become more palatable, meats become tender, and foods spoil less easily.

Heat is used to change the structure of proteins in many ways. When making cheese, curds are heated. This causes the proteins, which were already denatured by acid and the rennet enzyme, to bunch closer together and expel the whey to make the cheese firm and less likely to spoil.

In making custard, egg proteins are heated to make a firm gel, but we try not to overheat them. Overheating would cause the protein to contract and expel too much moisture, turning the custard into sweet scrambled eggs.

In making bread with milk, recipes sometimes tell you to scald the milk first. This denatures a protein in the whey that interferes with the volume of the loaf as it rises and bakes. Of course, overheating the milk scorches it, welding the milk proteins to the bottom of the pan, making it very difficult to scrub clean. Pasteurization and scalding perform the same functions, and many recipes that call for scalded milk were developed before most milk was pasteurized.

Volume Reducing and Drying

Sometimes foods are heated simply to drive off extra water to thicken the sauce, soup, or stew. In the process you get various reactions that change the flavor, color, and texture of the food, for the better or for the worse, but the main goal is reduction of volume and concentration of flavor.

Heating is a fast way of drying foods. Sliced apples dried in a 250°F (121°C) oven become crisp and light and are much less likely to spoil. Actually cooking the apples using a higher heat destroys the cell walls and some of the nutrients, while drying slowly without heat allows time for enzymes and microbes to spoil the food.

Flavor Producing

Heating food seldom makes it more sweet, sour, or salty. The other sensors on the tongue—those for bitter and savory tastes—are more often involved, especially if the food is burned. Heating sucrose can cause it to break down into glucose and fructose, and that combination is sweeter than the original sucrose. However, reactions like that are not the general rule.

For the sensors in the nose, however, heating is more often a benefit, as volatile molecules are freed and aromas are created and released. Maillard reactions and caramelization produce both bitter tastes and many volatile aroma molecules.

Heating also causes chemical reactions that don't happen at lower temperatures, or happen much more slowly. Many of those chemical reactions create flavor molecules. We have already looked into the most famous reactions—the Maillard reaction and caramelization—but other effects also create flavor when foods are heated.

Fats release flavor when they melt. Fats with higher melting temperatures retain volatile flavor and aroma molecules until the temperature exceeds their melting point. Vegetables release flavors and aromas when the pectins in the plant cell wall dissolve. Onions benefit from more than just the browning reactions. Heating deactivates the tear-producing molecules in them and makes them more pleasant to eat.

But not all heating produces beneficial results. Volatile flavor and aroma molecules can escape during heating and not be available when the food is served. Fats can oxidize, causing bad tastes or smells. Overcooking cabbage produces hydrogen sulfide, the foul-smelling gas that makes rotten eggs unpleasant.

Carcinogens

Muscles in cattle, pigs, birds, and fish contain a compound called phosphocreatine. At high temperatures—well above boiling, near 400°F (204°C) and above—it reacts with the amino acids in the proteins of the meat to create compounds called heterocyclic amines.

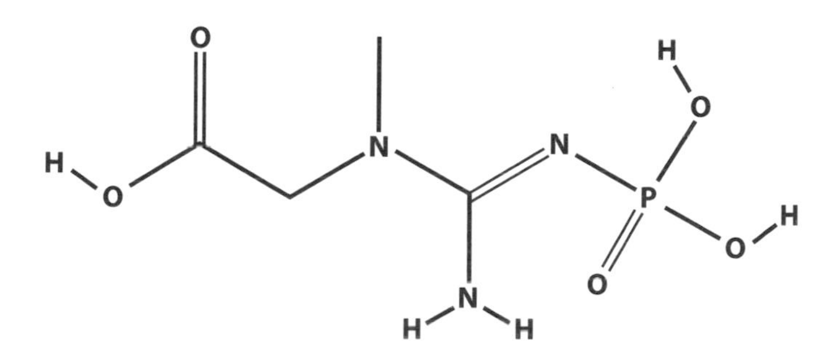

Phosphocreatine

Studies at the National Cancer Institute have shown that there is a link between people with stomach cancer and the consumption of cooked meats. These studies showed that the risk of stomach cancer was three times higher in people who cooked their meat well done than in people who cooked it medium rare. Other studies linked well-done meats to colorectal, pancreatic, and breast cancers.

The principal heterocyclic amines in overcooked muscle meats are *imidazoquinolines* (called IQ for short) and *imidazopyridines* (called IP for short). Five of the most prevalent are shown below. The word *heterocyclic* in this case means that they have a five-carbon ring attached to a six-carbon ring. The amine part is the nitrogen at the left that is attached to two hydrogens.

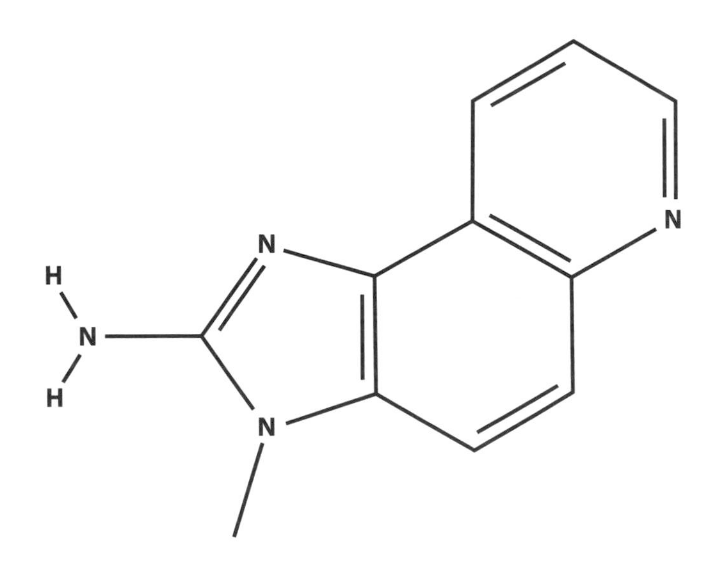

IQ: 3-methylimidazo(4,5-f)quinoline-2-amine

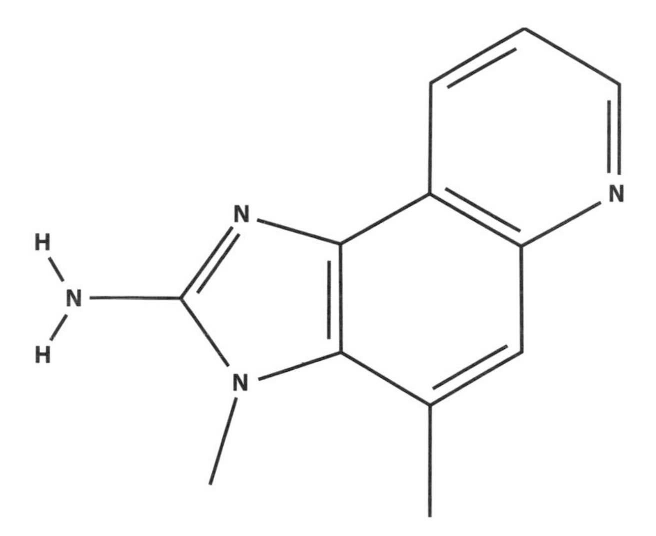

Methyl IQ: 2-amino-3,4-dimethylimidazo(4,5-f)quinoline

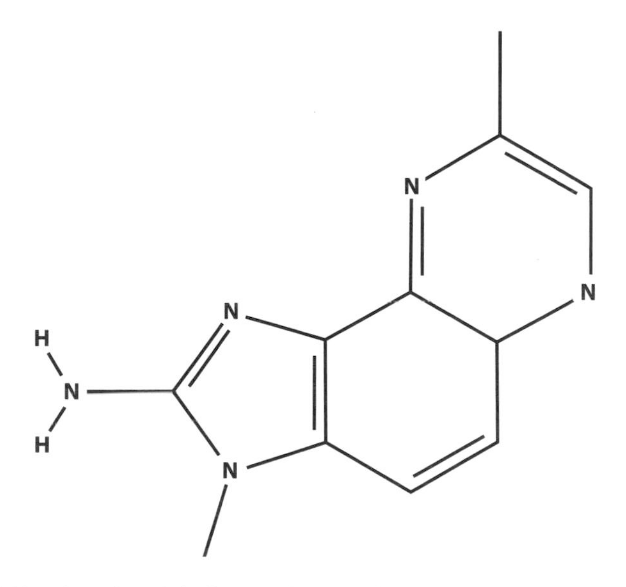

Methyl IQx: 2-amino-3,8-dimethylimidazo(4,5-f)quinoxaline

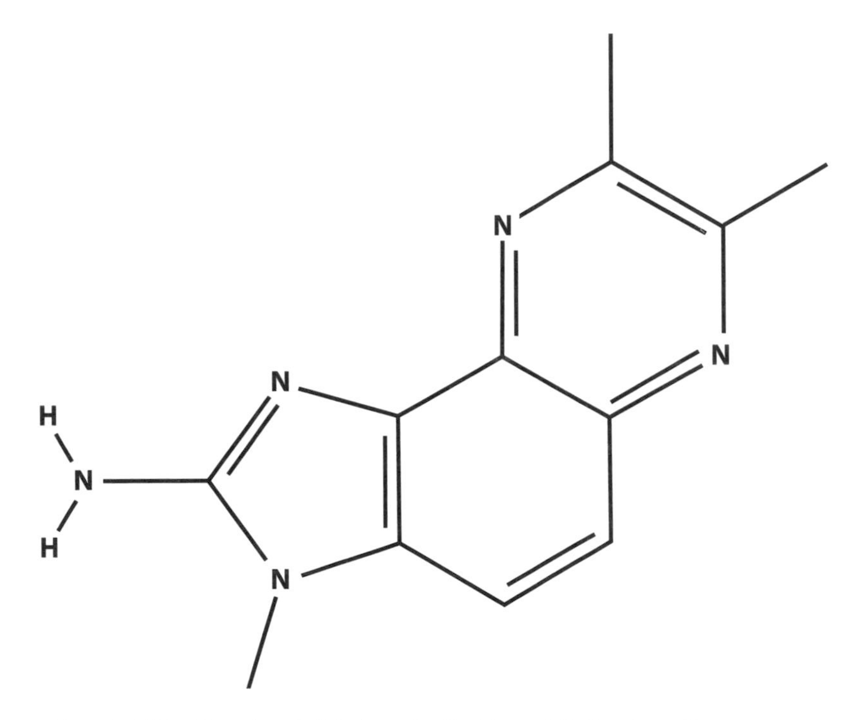

Dimethyl IQx: 2-amino-3,7,8-trimethylimidazo(4,5-f)quinoxaline

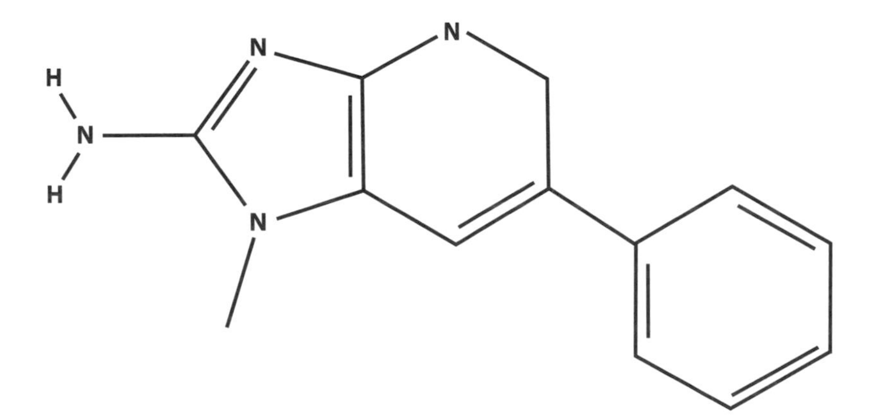

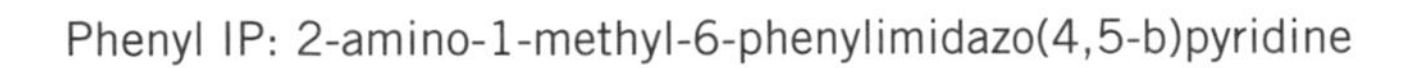
Phenyl IP: 2-amino-1-methyl-6-phenylimidazo(4,5-b)pyridine

Eating well-done meat increases your risk of cancer by 0.011 percent. For comparison, smoking cigarettes increases your risk of cancer by 7.9 percent.

Stewing the meat, where the temperature never rises above the boiling point of water, does not create these molecules. Microwaving the meat before grilling it reduces the precursors of these molecules, so much less is created during the high heat treatment on the grill. This reduces the cancer risk by 95 percent (to 0.00055 percent).

The same Maillard reactions that produce the flavors and aromas of heat-browned foods can also produce another type of carcinogen. Found especially in deep-fried potatoes such as french fries and potato chips, acrylamide has been shown to cause cancer in laboratory animals.

High-carbohydrate foods heated to above 250°F (121°C) are what create acrylamide, by reacting the amino acid asparagine with sugars.

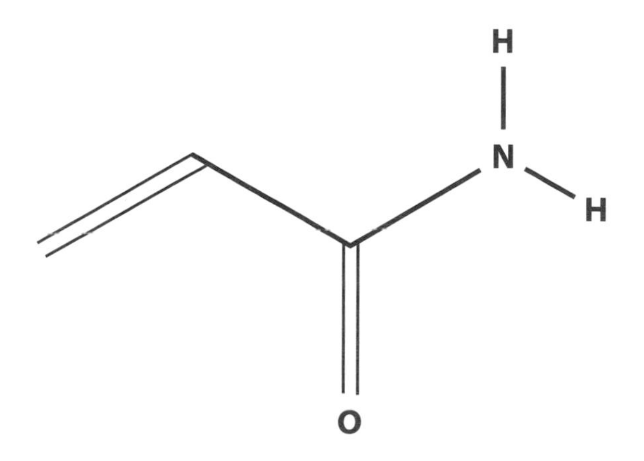

Acrylamide

However, studies in Swedish men found no connection to prostate and colorectal cancer when normal dietary amounts of acrylamide were eaten, and other studies showed no link to lung cancer in men, or endometrial cancer in women, and showed that it may in fact be protective against adenocarcinomas in women. One study of men and women in the Netherlands found some indications of a positive association between acrylamide and renal cancer risk.

So, while acrylamide is considered to be a probable human carcinogen, epidemiological studies have not shown any association between dietary acrylamide and cancer risk. Humans and rodents may have different absorption rates for this molecule, which might explain why laboratory animals get cancer from it while humans don't seem to.

Color Changes

One of the more obvious effects of heating foods is that they often change color. The browning of the Maillard reaction is one type of color change, but many others are also apparent.

Meat contains myoglobin, a red pigment that gives meat its color even when the red blood has been removed. When myoglobin is heated, it reacts with oxygen and turns brown, giving cooked meat the color we see when we slice into a well-done steak or pot roast.

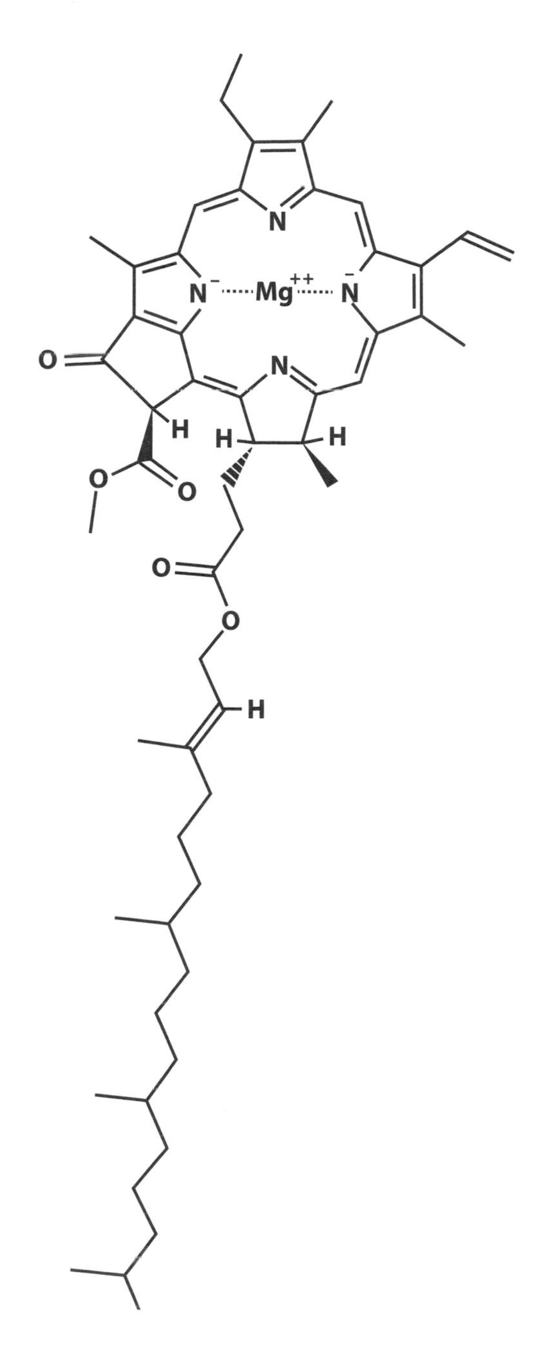

Chlorophyll A

Chlorophyll, the pigment that gives green vegetables their color, contains a central atom of magnesium. In cooking, that magnesium atom may be lost and replaced with a hydrogen atom. This changes the molecule enough that it turns a more grayish pale or yellow green. The color change also allows the red and orange carotenoid pigments to show through.

Plant cells have pockets of air in them that hide some of the green color of the chlorophyll. When heated, these gases expand and escape, and the result is the bright green color of lightly steamed broccoli. Cooking too long, or in an acid, will result in the loss of the magnesium and the dull color of overcooked vegetables.

Nutrition Changes

Cooking makes many nutrients in foods more available. Starches swell and soften, tough connective tissue turns into gelatin, and the pectins in plant cell walls soften into jelly.

Heating food can also protect some nutrients like ascorbic acid from being destroyed by natural plant enzymes. Heating can also destroy some antinutritional factors such as tannins in food and thus make the food more nutritious.

Heating can also destroy nutrients. Thiamin decomposes when heated in alkaline solutions, and eggs are slightly alkaline and become more alkaline with age. Boiling an egg results in an average loss of 15 percent of the thiamin in the raw egg.

And the Maillard reactions make the amino acids in proteins unusable as protein building blocks, although in general only a tiny portion of the protein in something like a steak participates in the reactions at the surface.

Leavening

Gases expand when heated. This is an important effect in the baking of any foam, such as cakes and breads. Cooks call the extra rising that occurs in the first few minutes of baking "oven spring."

Some baked goods such as popovers get all of their leavening when the water in the batter turns to steam, and the steam continues to expand with extra heat.

Another effect of heat comes into play in "double acting" baking powders. These powders are made from baking soda and two different powdered acids. One acid reacts immediately when water is added, and the acid then combines with the baking soda to create carbon dioxide gas. But the second acidic powder, sodium aluminum sulfate, reacts slowly at room temperature, but reacts much more during baking. This allows more time for a batter to be mixed or blended in the kitchen before baking. It ensures that the gases lost as the batter is prepared are replaced as it cooks in the oven.

12

ACIDS AND BASES

Acids are what make foods taste sour. In fact, the name comes from the Latin word for sour, *acidus*. Bases are substances that neutralize acids.

An acid is any molecule that can easily lose a hydrogen ion. A base is a molecule that accepts a hydrogen ion. Since a hydrogen ion is just a simple proton, acids are sometimes referred to as proton donors. Protons don't exist in the free state in water—the protons from the acid combine with the water to make H_3O^+, called a hydronium ion.

Water itself is both an acid and a base. Any two molecules in water always have a small chance—one in 10 million—of spontaneously converting into a hydronium ion (H+) and a hydroxide ion (OH–) when one of them donates a proton to the other.

Because a proton moves from one molecule to another, when an acid reacts with something, the result is generally two ions, one with a positive charge, and the other with a negative charge. When the water is removed (say, by evaporation), the two ions combine to form a salt. The familiar table salt NaCl forms when hydrochloric acid reacts with sodium hydroxide:

$$HCl + NaOH \rightarrow H^+ + Cl^- + Na^+ + OH^- \rightarrow H_2O + NaCl$$

The hydrochloric acid (HCl) dissociates into a proton (H^+) and a chloride ion (Cl^-), and the sodium hydroxide dissociates into a sodium ion (Na^+) and a hydroxide ion (OH^-). When the proton and the hydroxide ion combine and leave as water vapor, the resulting sodium and chlorine combine to form table salt.

Strong acids easily lose protons. Weak acids hold onto their protons a little better. Strong acids in water lose all of their protons to the water (making hydronium ions). Weak acids reach an equilibrium in which some of the molecules lose protons but others do not. Acids used in food, such as vinegar (acetic acid), soda water (carbonic acid), and lemon juice (citric acid), are weak acids.

Some acids can lose more than one proton. For example, carbonic acid can lose two protons, while citric and phosphoric acids can lose three.

Bases, called *alkalis* if an OH^- is involved, accept protons. Alkalis do this by donating hydroxide ions. These can accept protons from acids to make water. Lye (sodium hydroxide) is a familiar strong base. It is used to make soap and to clean drains by making soap out of the grease that has clogged the drain. The soap then dissolves and rinses away.

Ammonia is another familiar base, used for cleaning oils and grease from windows. Baking soda (sodium bicarbonate) is another base found in the kitchen, used because it reacts with acids to produce carbon dioxide gas.

To measure how acidic or basic a solution is, look at how many hydronium ions and hydroxide ions there are in the solution. If there are one in 10 million (as in pure water), we say the pH of the solution (think "percent Hydronium" as a memory device) is 7, because 10^7 is 10 million.

Smaller numbers for pH indicate a more acidic solution, and larger numbers indicate a more basic solution. Vinegar has a pH of about 2.4. Baking soda has a pH of about 9. Other examples are shown in the table below.

	$[H^+]$	pH	Common examples
Acids	1×10^{0}	0	Hydrochloric acid
	1×10^{-1}	1	Stomach acid
	1×10^{-2}	2	Lemon juice
	1×10^{-3}	3	Vinegar
	1×10^{-4}	4	Soda (carbonic acid)
	1×10^{-5}	5	Rainwater
	1×10^{-6}	6	Milk
Neutral	1×10^{-7}	7	Pure water
Bases	1×10^{-8}	8	Egg whites
	1×10^{-9}	9	Baking soda
	1×10^{-10}	10	Antacid
	1×10^{-11}	11	Ammonia
	1×10^{-12}	12	Quicklime (calcium hydroxide)
	1×10^{-13}	13	Drain cleaner
	1×10^{-14}	14	Lye (sodium hydroxide)

Effect of Acid and Heat on Sugar

Sucrose (table sugar) is a *disaccharide*, meaning it is two simple sugars, glucose and fructose, that have reacted in such a way that they join together, losing a molecule of water in the process. A reaction that produces water is called a *condensation reaction*. The fructose acts like an acid and donates a proton. The glucose acts like a base and donates a hydroxide ion. The proton and the hydroxide ion combine to form water, and the two simple sugars combine to form sucrose.

The reverse reaction, called *hydrolysis*, occurs when a water molecule is added to a molecule to break it into two parts. Hydrolysis of sucrose in water happens very slowly all by itself. But if an acid is added, it acts like a catalyst, promoting a faster reaction but not getting used up in the process. Heating up the solution makes the reaction go even faster.

The result of heating sucrose in water with a little lemon juice or vinegar in it is that much of the sucrose is converted into the two simple monosaccharides. Since fructose is a lot sweeter than sucrose, the result is a sweeter solution, even though glucose is not quite as sweet as sucrose. Since the acid is not used up, the solution is also a little tart, but that can be fixed by adding a weak base, such as egg whites or baking soda. If there are proteins in the solution, they can also react with the acid to neutralize it.

Effect of Acid on Proteins

You have seen the effects of acid on proteins in making yogurt (page 141). The casein proteins in milk stay in solution because they have water-loving protein strands on the outside of the micelles (clumps of protein) and calcium phosphate inside holding the micelles together. But at a pH of 4.6 (less acid than vinegar, but still quite sour), the calcium phosphate dissolves and the proteins denature, causing the micelles to clump together and form a gel.

While many cheeses are made with enzymes like rennet, which snip off the ends of the water-loving strands in the micelles to get them to clump together, other cheeses are made using acid.

Cheeses such as the Indian paneer, or the Italian ricotta, are made using acid (or acid and heat together) to coagulate the proteins in the milk. Since acid coagulates the whey proteins as well as the casein proteins, the resulting cheese does not melt when fried or baked, and the yield from a gallon of milk is higher (the whey proteins hold water better than casein proteins do). The resulting curd is not as firm as a rennet cheese, due to the different way the proteins bind together as they coagulate.

Cooking with Acid

Acids are used to denature other proteins in cooking. Fish is "cooked" in lime juice in traditional ceviche recipes, and eggs are "hard boiled" by pickling in vinegar.

Most acid cooking is done with seafood, where the meat does not have as much of the tough collagen connective tissue as beef does. Acid cooking does not soften connective tissue in the same way heat does (by making gelatin out of it). But the acid does denature the proteins in the fish or shellfish, giving them a cooked mouthfeel and a cooked appearance. The acid also does not do as good a job of sterilizing the meat, so cleanliness and freshness are very important.

Acids are also widely used in marinades. Vinegar, tomato juice, citrus juices, and yogurt are often used to denature the outer proteins on the meat, so they open up and absorb the other flavor elements in the marinade. Marinades do not penetrate very far into the meat, so often the meat is pierced many times with a fork to get as much surface area exposed to the liquid as possible.

Cooking with Alkali

As with acids, alkalis can be used to "cook" foods. However, if the food contains fats, the alkali turns them into soap, which people often object to eating. Something about the taste.

But since soap washes away easily (along with any water-soluble vitamins and minerals), washing the result can make it palatable.

A notable use of alkali in cooking is the process known as *nixtamalization*. This wonderful word refers to the practice of cooking maize (the name for what we Americans call corn) in alkali. The name comes from the Aztec words *nixtli* (for ashes) and *tamalli* (for uncooked dough made from corn). The ashes from a cooking fire are alkaline, containing potassium hydroxide. In modern corn processing, calcium hydroxide (lime water) is used instead.

Boiling corn kernels in alkali softens the hard outer hull. The starches absorb water and swell, and then form a gel. Enzymes from the germ of the seed are released and act on the starches and proteins, improving the workability of the resulting dough.

The cell walls of plants are made of cellulose and pectin, which dissolve in hot alkali solutions. But the hot alkali also denatures the corn proteins, making them more available to human digestion.

Nutritionally, the corn is further improved (from the perspective of human digestion) by freeing the niacin that is otherwise bound to proteins in the corn. When the proteins are denatured, the niacin can be digested. Other animals can make their own niacin from the amino acid tryptophan, or they can digest the proteins that hold onto the niacin, and get niacin from corn that way. But humans have lost that ability, and we need to cook the corn in alkali to free those nutrients.

Minerals in the lime water or in the ashes are also absorbed by the corn in the process of steeping it in the liquid after boiling. This increases the calcium content of the dough, along with increasing the iron, zinc, potassium, and copper content.

There is fat (corn oil) in the kernels, and some of this turns to soap in the process. For this reason, the liquid remaining is usually discarded, and sometimes the dough goes through an additional rinse.

Sodium bicarbonate (baking soda) is probably the most familiar alkali used in cooking. It is actually the salt of a strong alkali and a weak acid, so that it acts as a weak alkali. The strong alkali is sodium hydroxide (lye), and the weak acid is carbonic acid (soda water). Combining the two together first creates a related compound, sodium carbonate (washing soda).

$$CO_2 + 2NaOH \rightarrow Na_2CO_3 + H_2O$$

Combining sodium carbonate with more carbonic acid makes sodium bicarbonate.

$$Na_2CO_3 + CO_2 + H_2O \rightarrow 2\ NaHCO_3$$

The two molecules are similar. The bicarbonate has a hydrogen where the carbonate has a second sodium. Thus another name for the bicarbonate is sodium hydrogen carbonate.

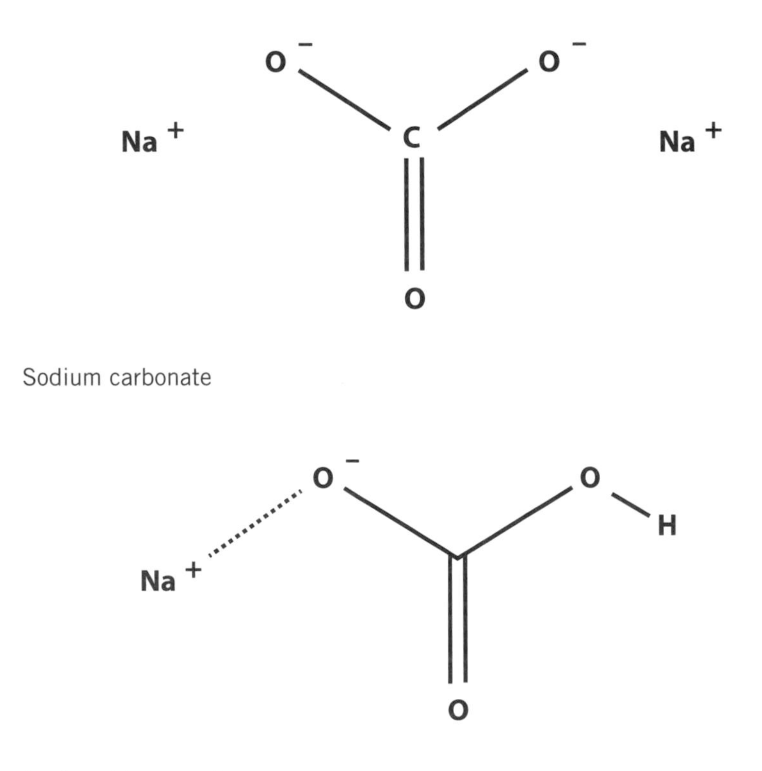

Sodium carbonate

Sodium bicarbonate

Sodium bicarbonate is made by bubbling carbon dioxide gas into a solution of sodium carbonate. But the sodium carbonate used is not generally made using lye and soda water. It is either mined in the form of a mineral called trona (a mixture of sodium carbonate and sodium bicarbonate) or it is made by the Solvay process, which is more complicated (saltwater, ammonia gas, and calcium carbonate are used in a multistage process that turns out to be cheaper than the simpler reaction).

Most of the uses of baking soda in the kitchen involve its alkaline nature, although it is also used as a mild abrasive in cleansers and toothpaste. It reacts with acids to release carbon dioxide for leavening baked goods, it reacts with acids in the stomach to relieve indigestion, and it reacts with acid molecules in the air to freshen refrigerators.

At 158°F (70°C) sodium bicarbonate breaks down into sodium carbonate, water vapor, and carbon dioxide gas and thus can be used to make foamed candy by adding it to a very hot syrup. For the same reason, it can be used as a fire suppressor in fire extinguishers or simply by dumping a box of it on a kitchen fire.

pH-Sensitive Colors

Many colored molecules react with acids and bases in ways that change their colors. Chemists use this color change to indicate how acidic or basic a solution is. Litmus paper is one such indicator, but there are a large number of other indicators available.

A very common class of indicators is the group of pigments called *anthocyanins* (from the Greek roots for "flower" and "blue"). These molecules reflect red light in acid solutions and blue light in basic solutions. Many of the colored fruits, leaves, and flowers you encounter owe their color to anthocyanins and to the levels of acid in the plant. Examples are the red of apple skins and cherries, the red or purple of red cabbage leaves and eggplant skins, the blue of blueberries, the red of red wines, the purple of purple corn, the blues and reds of pansies and violets, and many more.

A common high school chemistry demonstration is making a broth from red cabbage and then showing how it turns red when vinegar is added and blue when baking soda is added. The same effect can be seen when washing a glass that has had red wine or grape juice in it. Since most tap water is neutral to slightly alkaline, the red wine will turn blue when the water is added and the acid is diluted or neutralized.

Sour Sensing

Acidic foods taste sour, unless there is sugar along with the acid. Lemonade is made by taking acidic fruit juice and adding sugar. Most lemons are actually no more acidic than orange juice. The orange just has more natural sugar than lemons do.

What this tells us is that the tongue and the brain give us different information about acid, depending on how sweet the food is. Acidic things that are not sweet are often not good to eat. They may be spoiled food, polluted water, or some other thing we should not be putting in our mouth.

But fruits are often both sweet and sour at the same time. Our sense of taste has adapted to this situation, and we find the combination pleasant, an indicator of something good to eat. Sweet things contain much needed energy, and not eating them because of their acidity is not a good survival mechanism.

Recipe

Lemonade with Chameleon Eggs

A popular topic in cooking these days is a neat trick for making little balls of gel that look like caviar or salmon eggs. They are made out of a gelling agent called sodium alginate. This compound is extracted from kelp, and when added to water it makes a thick syrup. When calcium ions are added to this syrup, they exchange with the sodium, and a tough gel forms. Using an eyedropper or a pipette to let the syrup fall into a solution of a calcium salt a drop at a time makes wonderful little balls of gel that look like fish eggs or frog eggs.

Now take that trick a step further and make these little eggs slowly change color from blue to red when they are stirred into a glass of lemonade. You can do this by using the anthocyanin pigments from the skins of red grapes as a pH indicator. Start the eggs out at a neutral to slightly alkaline pH, and let the acid in the lemonade bring the acidity up until the anthocyanins turn from blue to red.

You can buy sodium alginate at health food stores for about $4 for a 2-ounce jar, which is six times as much as this recipe needs. Or you can buy it on the Internet for about

$40 a pound; it will keep for years. I use a tablespoon of powdered sodium alginate, which is about 9 grams (a third of an ounce).

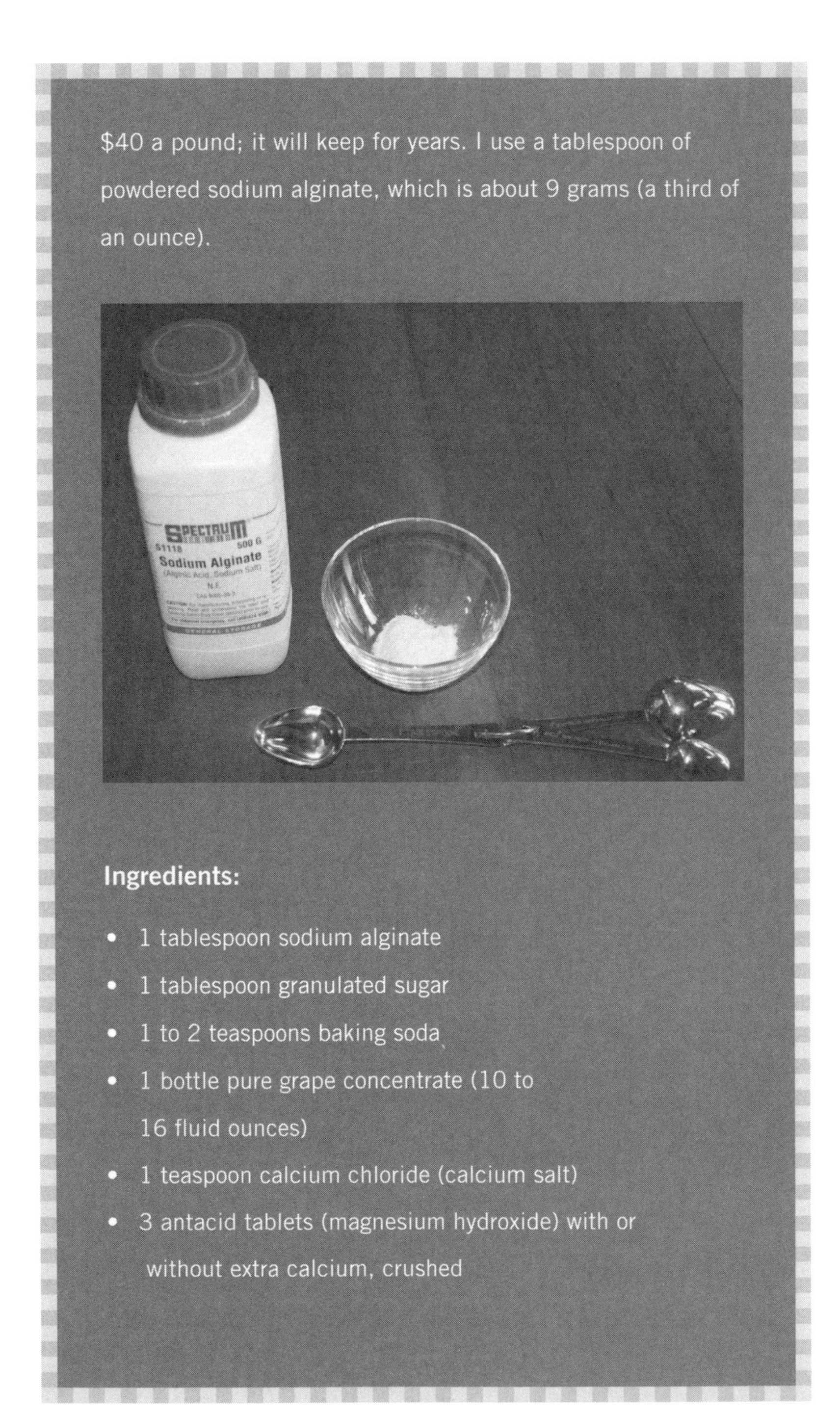

Ingredients:

- 1 tablespoon sodium alginate
- 1 tablespoon granulated sugar
- 1 to 2 teaspoons baking soda
- 1 bottle pure grape concentrate (10 to 16 fluid ounces)
- 1 teaspoon calcium chloride (calcium salt)
- 3 antacid tablets (magnesium hydroxide) with or without extra calcium, crushed

Supplies:

- Jar
- Spoon
- 12- to 16-ounce glass
- Eyedropper or pipette
- Slotted spoon or strainer

If the grape juice is frozen, let it thaw until it is liquid. You want to make sure the juice has not had extra acid added to it. If you can't find concentrate, you can use regular grape juice. The result will still be good, but it will not burst on the tongue in quite the same way.

Next, mix the sodium alginate and the sugar together. This will make it a lot easier to dissolve the sodium alginate in the liquid. Since the alginate will start to gel as soon as it gets wet, it will form clumps and lumps that will take a lot of stirring and mashing to dissolve. The sugar helps by letting the liquid get into the mass of powder.

Now you want to neutralize the acid in the grape juice, without going so far that you can taste the baking soda. The alginate will not dissolve in an acid liquid, and you want the grape juice to be quite blue in color, as it gets when it has a neutral pH. Add the baking soda to the grape juice ⅛ teaspoon at a time. Stir until all the bubbles have disappeared before adding the next ⅛ teaspoon. At some point you will notice that fewer bubbles came up from that last bit of baking soda. This is the time to stop. You don't need to go all the way to neutral. Be patient. Waiting for the bubbles to go away can take a while. Don't be tempted to use the antacid tablets instead of the baking soda—they will cause the alginate to solidify immediately.

Put the alginate powder and sugar in a jar, pour the liquid grape concentrate on top, and begin to stir. Use a jar so that once the main lumps have been broken up with the spoon, you can put a lid on it and shake it. Even with all this trouble, you will have to let the mixture sit for an hour or two to let the gel fully hydrate. Overnight is even better. Plan ahead; do this part the day before you want to impress your friends.

When the powder has all dissolved into a somewhat thick syrup, dissolve the calcium chloride and crushed antacid tablets in a large glass of water (12 to 16 ounces). Stir well, and leave the solution swirling for the next step.

With the eyedropper or pipette, suck up some of the grape syrup and hold it above the antacid solution so that single drops can be squeezed out to fall into the antacid. But don't let the eyedropper touch the antacid, or the syrup will begin to gel in the eyedropper and clog it up.

The drops might be red or purple as they form spheres in the antacid solution. Leave them in the solution until they turn blue. Then you can remove them with a slotted spoon, or pour the solution through a strainer into another glass. The chameleon eggs will remain in the strainer.

You will want a teaspoon of eggs for each glass of lemonade. This will take a few minutes to make using the eyedropper. If you are planning a big party, you may want to purchase a portable showerhead with a hose at a hardware store and make a lot of drops at a time.

Serve the eggs in a spoon next to the glass of lemonade. Your guests can then stir them into the lemonade and watch them slowly turn red during the meal.

Putting them in a lemon-lime soda also works—not only are they easier to see, but the carbonation makes the eggs float up and down as the bubbles first attach to the egg and make it rise, and then pop when it reaches the top, letting them fall back down, like little submarines.

Chameleon eggs will keep for a long time in a jar in the refrigerator.

If you want a quicker treat, without the color change, try adding alginate to strawberry syrup and dropping that a drop at a time into a calcium chloride solution. You get pink caviar that tastes like strawberry syrup. Serve it on a spoon, or spread it on crackers with cream cheese.

13

OXIDATION AND REDUCTION

In acids and bases, protons (hydrogen ions) can be transferred between molecules. Those protons are positively charged subatomic particles. The most familiar negatively charged subatomic particle is the electron. When electrons are transferred between molecules, the process is called *oxidation*.

Something is said to be *oxidized* when it loses electrons, and something is *reduced* when it gains electrons. It may seem backward to think of gaining electrons as being "reduced," but the name is a historical accident. Benjamin Franklin chose the names for positive and negative charges before the discovery of the electron. Since electrons have negative charges, gaining an electron makes the oxidation number smaller, because you are adding a negative number.

As a proton donor is called an acid, an electron donor is a reducing agent. An electron acceptor is an oxidizing agent. (This simplified picture of oxidation is enough for this kitchen discussion, but as with most things, it can get more complicated as you drill down deeper.)

When you burn a candle, the oxidizing agent is the oxygen in the air, which steals electrons from the carbon and hydrogen in the candle wax. The result is oxidized carbon and hydrogen, known as carbon dioxide and water, respectively.

To cook on a gas stove, you don't need to know much more than that. But much of the oxidation in the kitchen happens more slowly. Iron rusts, apple slices brown, oils and fats get rancid, and wine turns into vinegar. We also breathe in oxygen and breathe out carbon dioxide and water vapor.

Apples, Avocados, and Lemon Juice

Plants produce an enzyme called polyphenol oxidase. An *oxidase* is any enzyme that oxidizes something. In this case, the thing being oxidized is a polyphenol.

You read about polyphenols earlier; they just weren't grouped together under that name. Vanillin is a polyphenol. Tannins are polyphenols. Anthocyanins are polyphenols. Polyphenols from wine (resveratrol) and green tea (EGCG, or epigallocatechin gallate) are sold in health food stores. They are part of a larger class of compounds called antioxidants, because they are good at combining with harmful oxidizers. Their health benefits probably come from their other biological activities, however, as there is doubt that they act as antioxidants in the human body when consumed.

There are colorless polyphenols in apples, and there is polyphenol oxidase in apples. These two molecules are normally kept apart, but when you slice into an apple, you break open cell walls, which allows the two to mix and react.

As the enzyme combines the polyphenols with oxygen from the air, it produces quinones, which then combine (polymerize) to make dark colored molecules called melanins.

In an apple, one of the polyphenols is chlorogenic acid:

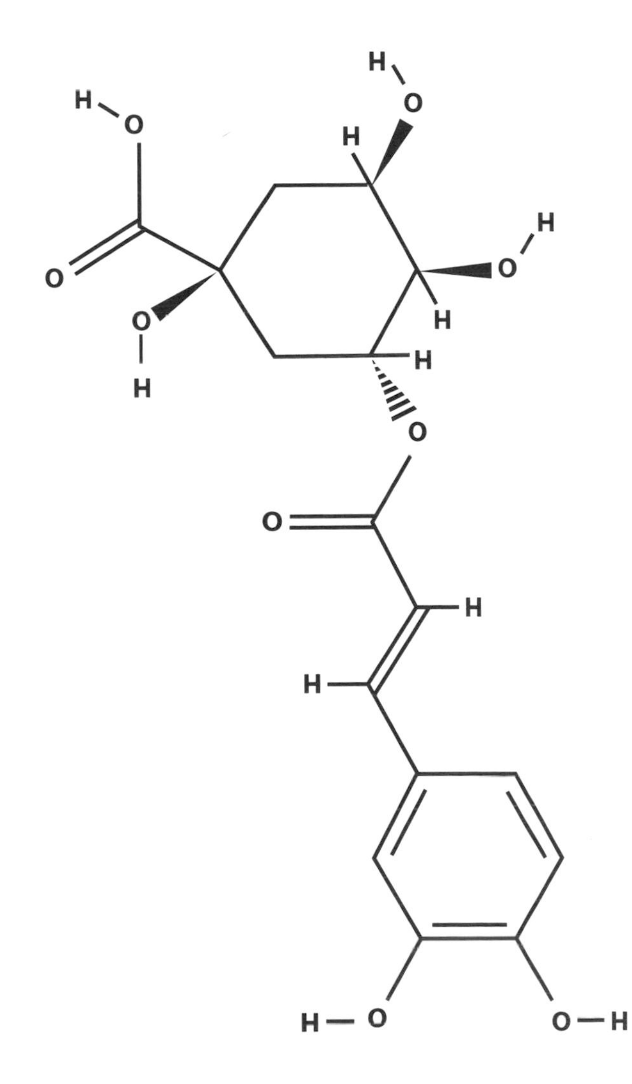

The phenol part is the hexagon on the bottom, with the two hydroxyl groups attached to it. The polyphenol oxidase enzyme removes the hydrogens from those to make a molecule called a quinone. A series of reactions then converts the quinone into melanin:

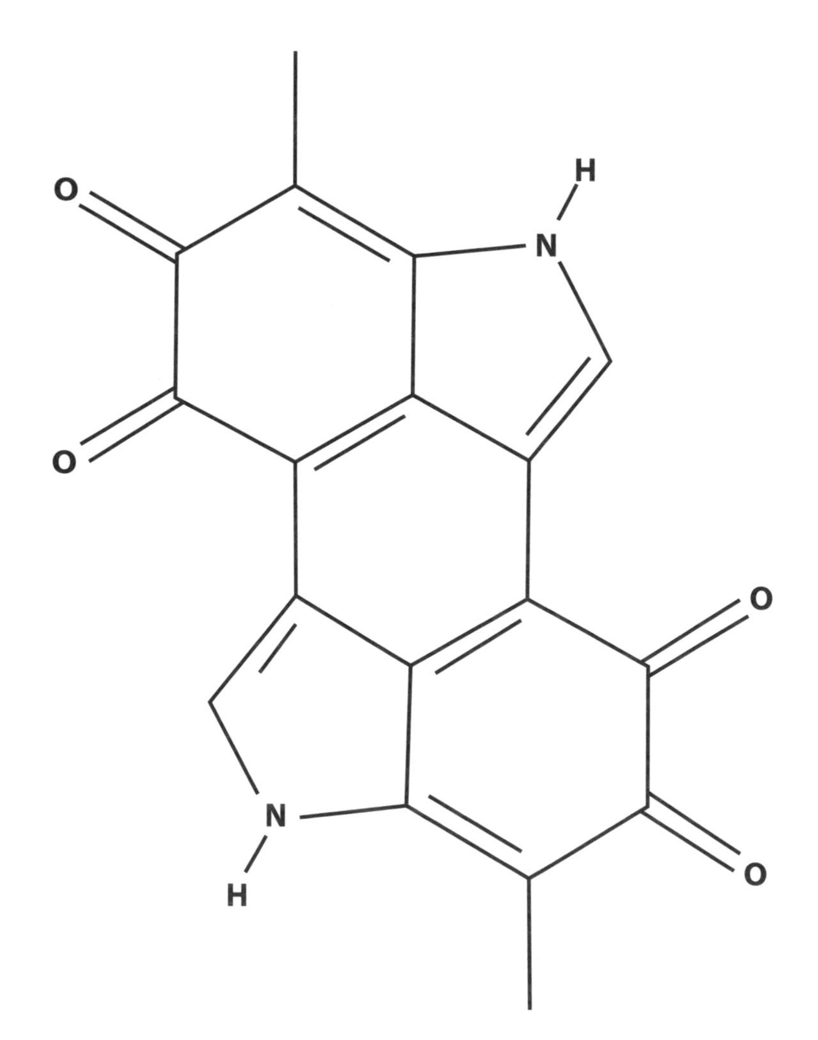

Melanins are the molecules responsible for the brown in brown hair, the red in red hair, and the brown of a suntan and of freckles. They are found in the ink of squids and octopi. And in the cut apple, they make the unappetizing brown stains on the surface, or brown bruises below the surface.

There are a number of ways to interfere with these complex chemical reactions and thus prevent the fruit from browning. Enzymes are proteins, and they can be denatured with heat until they no longer function. *Blanching* fruit by dipping it in boiling water will prevent browning.

The polyphenol oxidase enzyme also only works in a limited range of acidity. Making the juice on the apple slice too acidic or too basic will prevent the enzyme from performing its function. A little citric acid sprinkled on the cut will do the trick.

Enzymes work more slowly, or not at all, when the temperature is reduced. Refrigeration or freezing will slow or prevent browning.

Like most enzymes, polyphenol oxidase needs water in order to work. Dehydration will prevent browning. However, dehydration takes time, and the browning reaction is quick, so while the dehydration is happening, you can use yet another trick: simply prevent oxygen from the air from getting to the cut fruit. You could also freeze-dry the apple slices.

You could denature the enzyme with X-rays (irradiation) or electron beams. High pressure can also denature the enzyme.

Or you could remove the oxygen by using a more powerful reducing agent than the polyphenols. Ascorbic acid (vitamin C) can be sprinkled on the cut fruit. It will combine with the oxygen and make it unavailable to the enzyme. It can also make the juice more acidic at the same time, so it has a double action. And, of course, you have just improved the nutrition of the fruit by adding a vitamin.

Other reducing agents used are sulfites, which are often added to cut fruit before drying in the sun.

Vinegar from Wine

Vinegar is made from wine that has been left open to the air. The ethanol in the wine oxidizes into acetic acid. The oxygen attaches to the central carbon atom:

Ethanol (above) becomes acetic acid (below)

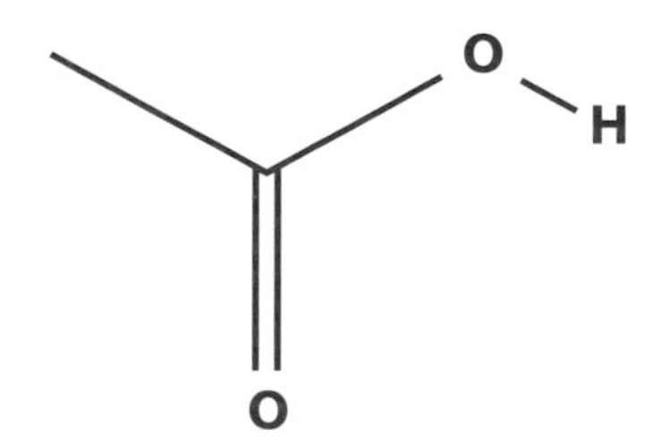

But the ethanol doesn't simply "burn" or "rust" into acetic acid. The transformation is done with enzymes, and the enzymes are produced by a bacterium called *Acetobacter aceti*.

The bacterium, like most organisms, has a preferred environment. It grows best at around 80°F to 85°F (27°C to 29°C), with no more than 15 percent alcohol, with 9 percent to 12 percent being optimal. It also needs oxygen, unlike the yeast that made the alcohol in the wine.

As the bacteria grow, they form a mat of cellulose and other polysaccharides at the bottom of the wine. This mat is called "mother of vinegar," because adding it to wine provides the starter culture of bacteria to make the vinegar.

Vinegars can be made from any fermented alcoholic beverage, and often are. Malt vinegar is made from ale or beer, for example. Fruit vinegars, rice vinegar, honey vinegar, and vinegar from sugarcane are popular in many places around the globe.

Oxidation of Oils and Fats

Oils and fats react with oxygen in ways that are sometimes beneficial and sometimes undesirable. When cooking oils react with oxygen, they can form compounds that taste bad; they're rancid. But when oils react with oxygen and polymerize into tough insoluble films, they're used in paints and coatings for wood products.

Chemistry Lesson

How Oxygen Forms Molecular Bonds

An oxygen atom has six electrons in its outer electron shell, but eight electrons can fit in that shell. If two more electrons can be borrowed from another atom to form a molecule, the molecule will have a lower energy than the two separate atoms had, resulting in a strong covalent bond. To break the bond would require adding the energy back. In the drawing below, two oxygen atoms are shown, with the empty slots for two more electrons shown in gray.

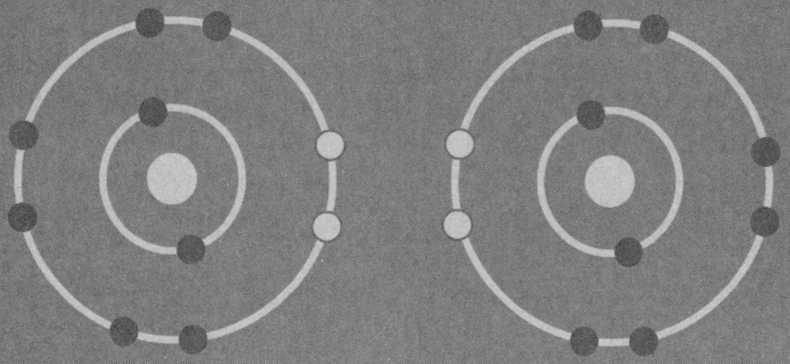

Two oxygen atoms can form two covalent bonds with one another. This lets two oxygen atoms fit together so that they share the outer electron shell that has eight slots for electrons. The shared electrons fall into the holes in the outer shell and fill them up, locking the two atoms together. This is called a *double bond*.

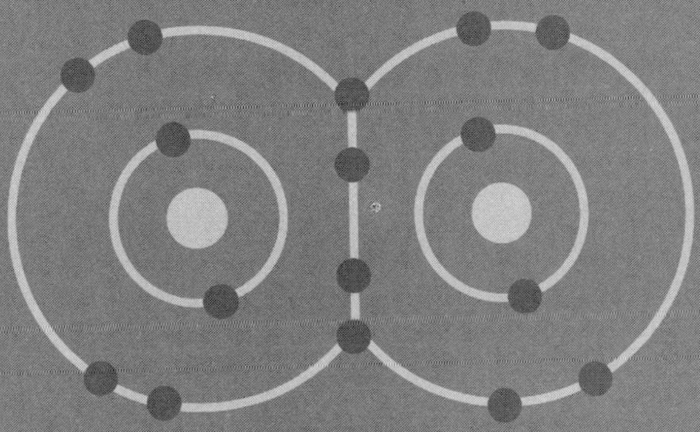

Oxygen molecules can also form with a *single bond* between the atoms. This leaves one lone electron on each oxygen atom.

There is a rule in quantum mechanics called the Pauli Exclusion Principle, which accounts for why atoms take up space. It limits how many electrons can exist in the different orbitals of the atom. If there is no room in a low-energy orbital, the electron can't fall into a lower-energy state, and it stays in the higher-energy orbital. Two electrons can occupy the same orbital if they have opposite spins.

Free Radicals

When there is only a single bond between two oxygen atoms, the two lone electrons in each oxygen atom are not paired with an electron of the opposite spin. Such unpaired electrons are called *radicals* (sometimes free radicals) and are very reactive, since another electron from another atom can fall into the empty slot to pair with the lone electron and form a bond (and thus a new molecule).

Since the oxygen molecule with a single bond has two unpaired electrons, it is called a *diradical.*

The double bond between the two atoms in an oxygen molecule was broken into a single bond and two unpaired electrons. Bonds in other molecules can also be broken this way to form radicals. Single bonds between two atoms can be broken such that each remaining part has an unpaired free electron, so that two radicals are formed. They can escape to react with other molecules, or they can join back together, releasing energy.

Oils and fats oxidize by a process called a *radical chain reaction*. A photon of light can start the reaction by breaking a single bond between a hydrogen and a carbon in a fatty acid in the oil.

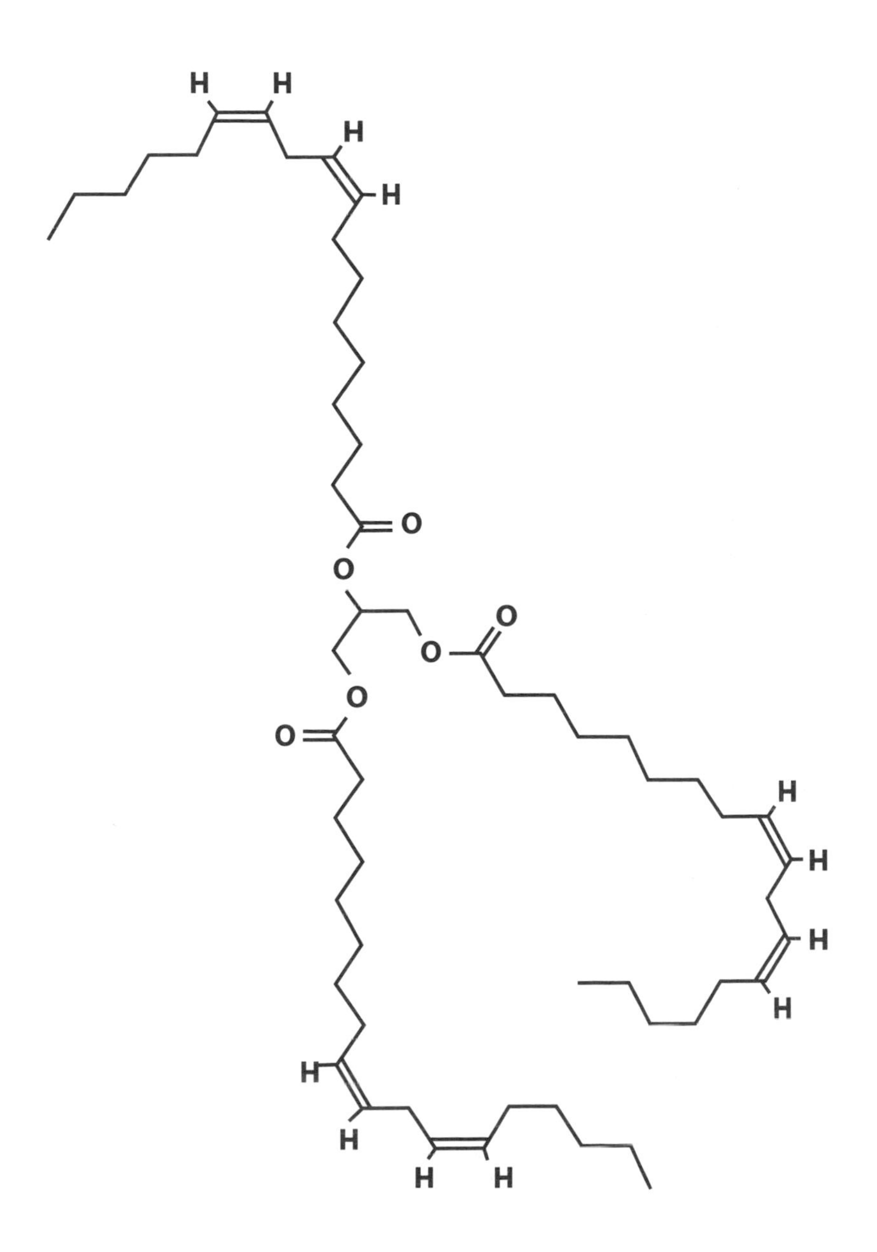

This results in a radical, where the carbon now has an unpaired electron.

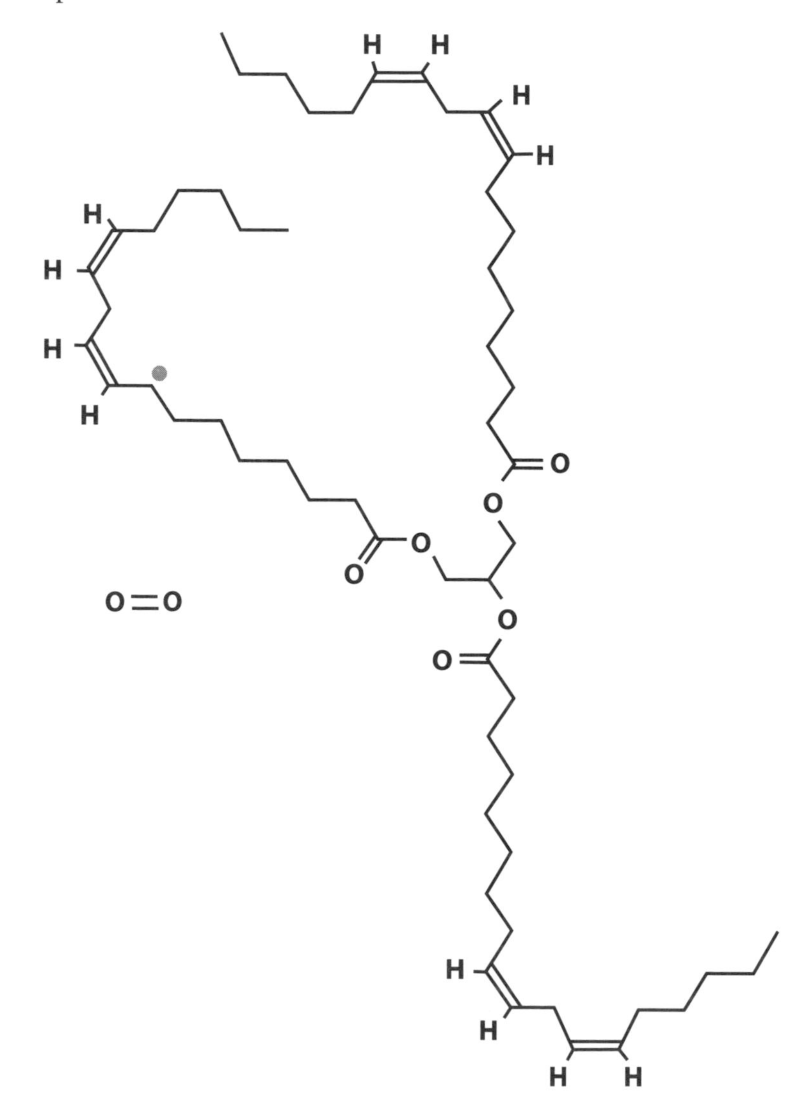

The unpaired electron then reacts with an oxygen molecule. It breaks the oxygen double bond and pairs with one of the resulting unpaired electrons, leaving a peroxide radical attached where the hydrogen had been.

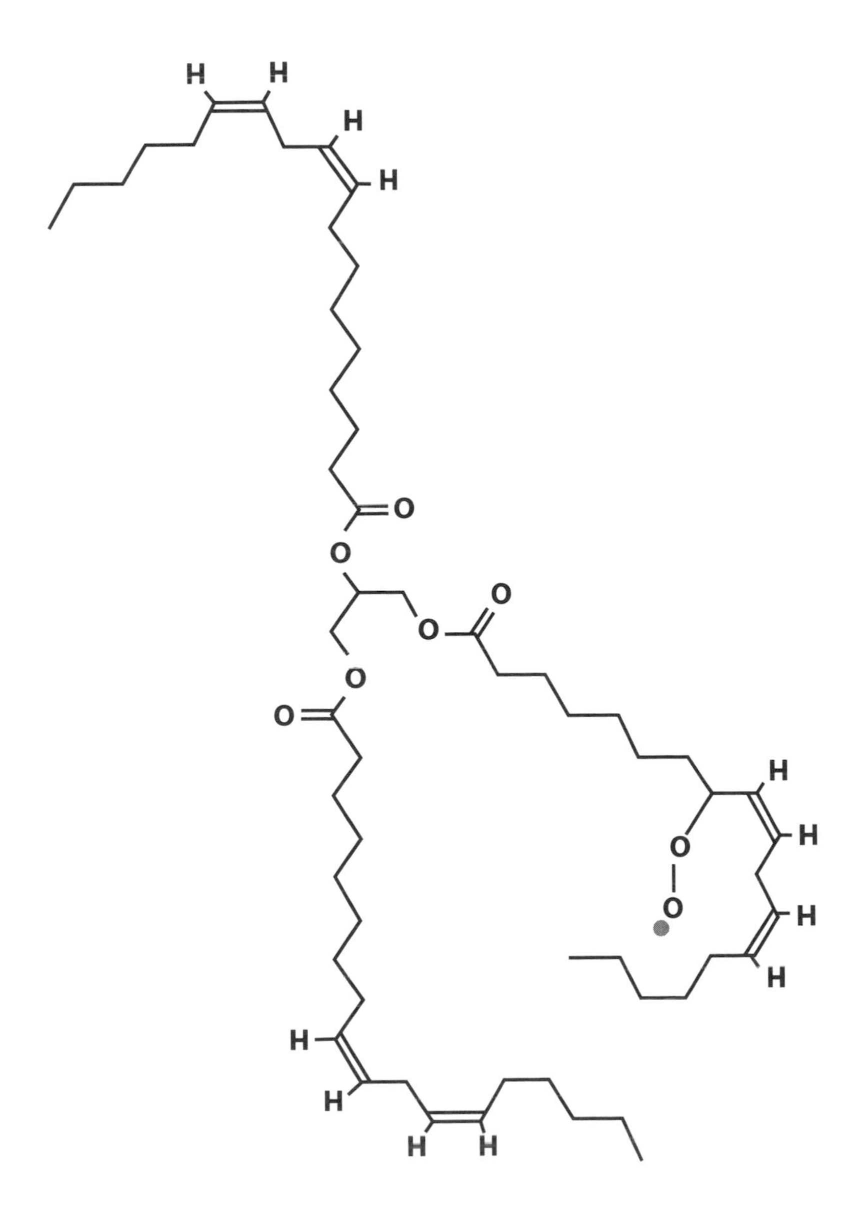

The remaining unpaired electron in the peroxide radical then steals a hydrogen from another linolenin molecule, which continues the chain reaction by bringing you back to where you started.

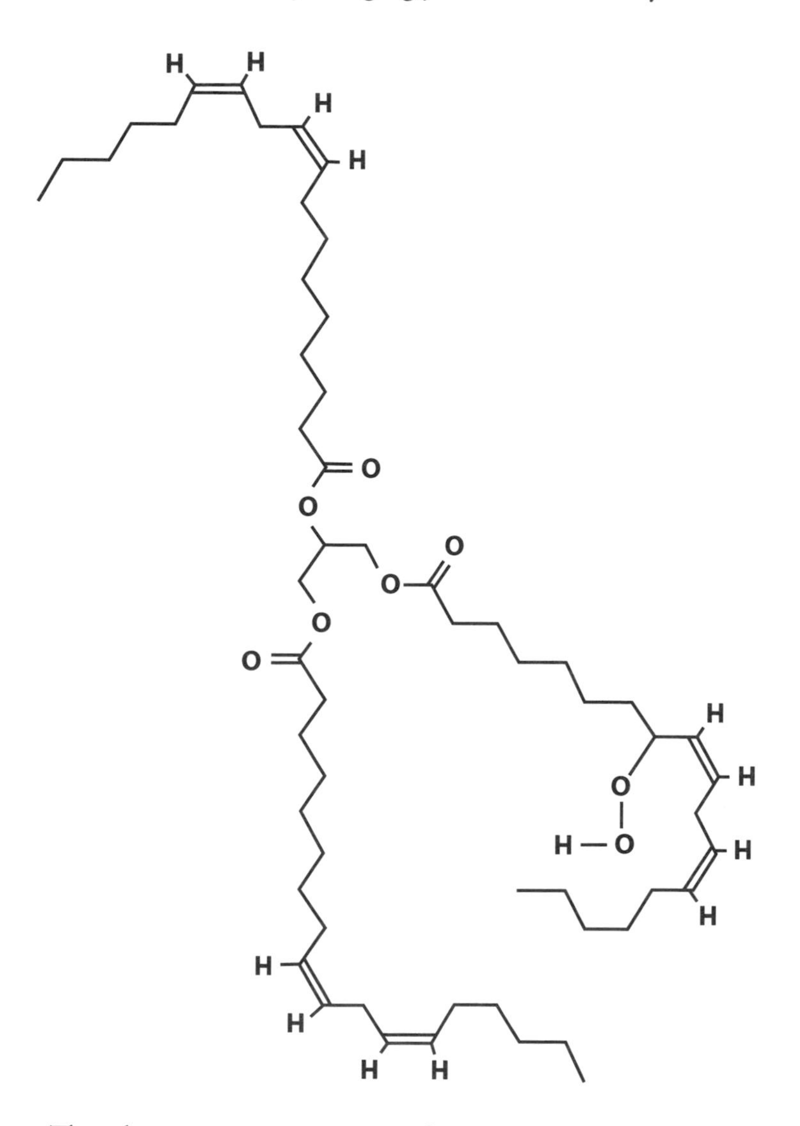

This chain reaction can go on for a long time, causing many of the oil molecules to be oxidized and become rancid. This is one reason why radicals are so damaging.

Eventually the radical meets up with another radical, and they cancel each other out by sharing electrons to fill out each other's pair. At this point the chain reaction stops.

So to keep oils from going rancid, you can either prevent the chain reaction from starting, by keeping the oils in the dark so no photons of light can create radicals, or you can slow down the reactions by keeping the oil cold in the refrigerator, or you can stop the reactions by using antioxidants to bind with the radicals before they can do too much damage.

Antioxidants

Some molecules have *stable radicals*. These are unpaired electrons that are either tucked in a crevice in the molecule where they aren't readily available or are next to a conjugated ring that stabilizes the unpaired electron and makes it less reactive. These molecules can stop the chain reaction by pairing up with the radical that would otherwise react with the oil.

There are many such molecules, called antioxidants. Alpha-tocopherol, or vitamin E, is one of them.

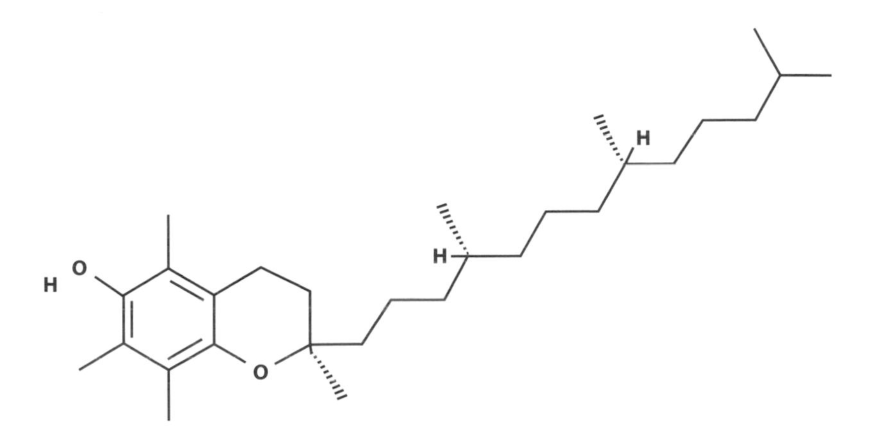

Another one is butylated hydroxytoluene, or BHT.

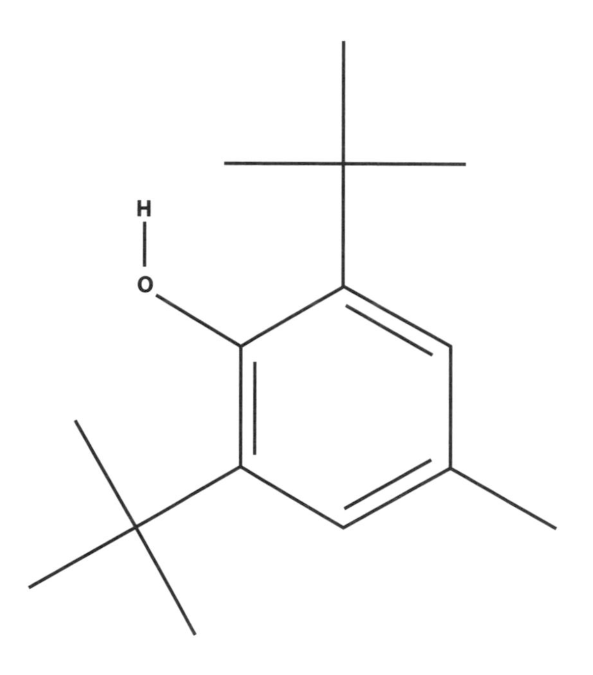

You have seen other molecules earlier with conjugated rings (depicted by hexagons with alternating double and single bonds). Anthocyanins are good antioxidants. In fact, many of the colored parts of plants owe their color to antioxidants.

❖ 14 ❖

BOILING, FREEZING, AND PRESSURE

It is fairly simple to lower the boiling point of water: just lower the pressure of the air pushing down on the water. This makes it easier for the water molecules to escape into the air. In fact, you can make water boil at room temperature if you lower the pressure enough.

An extreme version of this is the process of freeze-drying food. Under enough of a vacuum, the water molecules in ice will boil off, leaving the food dry. The food will have an airy crunch, since it does not shrink as much as when it is dried above the freezing point, and when air is let back into the chamber, it fills up the cavities where the ice was.

Altitude

Not everyone has a vacuum chamber yet (although they are getting quite inexpensive these days). But another way to lower the air pressure is to travel up a mountain.

Cooking foods at high altitudes is different from cooking at sea level in several ways. Water boils at a lower temperature, so cooking rice or eggs in boiling water takes longer. Leavened baked goods have several problems at high altitudes. They rise faster, so yeast breads double in volume before they can get the same flavor from the yeast as they do at sea level. Cooks adjust for this by punching the dough down an additional time and letting it rise again.

Cakes also rise too fast, so when the bubbles pop, the cake collapses. Cooks adjust for this by increasing the strength of the dough by adding more egg and more flour, and cooking the cake at a higher temperature for a shorter time so the dough sets before the bubbles pop.

Cakes also dry out faster at higher altitudes, both because the low pressure allows water to evaporate more quickly and because higher altitudes often have lower humidity than sea level locations. Cooks add a little more water to make up for this.

Adjusting the amount of leavening is also helpful, since a teaspoon of baking powder will make the same amount of gas as at sea level, but that gas will expand to a larger volume at higher altitude.

Shortening reduces the strength of the gluten molecules in the dough, so reducing the amount of shortening will help prevent the collapse of the bubbles. So will reducing the amount of sugar slightly, or replacing some of the sugar with nonfat powdered milk, which adds strength.

Raising the Boiling Point

Just as you can lower the boiling point of water, you can also raise its boiling point. Simply increasing the pressure will do this, which is how pressure cookers work. The water boils and steam is created, but it can't escape until it builds up enough pressure to lift the weight on the escape valve.

But you can also raise the boiling point of water by dissolving something in it that does not itself boil (is nonvolatile). This works by diluting the water so that fewer water molecules are at the surface and thus fewer can escape the liquid stage. This effect is referred to as reducing the vapor pressure of the water.

How much the substance raises the boiling point depends on how many particles of it (the solute) are dissolved in a given amount of water (the solvent). Some compounds break into more than one particle when they dissolve. Table salt breaks into a sodium ion and a chloride ion, so there are two particles in solution. Sugar stays as one particle. Calcium chloride breaks into three particles—a calcium ion and two chloride ions.

This comes into play when you try to calculate how much the boiling temperature will rise when you dissolve some amount of sugar or salt into water.

Adding a teaspoon of salt to a quart of water (or 10 grams of salt to 1 liter of water) will raise the boiling point by 0.31°F (0.17°C). That small amount is not noticeable when cooking. To make a noticeable difference, the amount of salt would have to be so much as to make the food completely inedible. In other words, adding salt to boiling water to make something cook faster just doesn't work in practice.

As you have seen earlier, sugar reacts with water and with itself to make larger molecules when it is heated. Also, sugar melts at a temperature not that much higher than water. For these reasons, sugar raises the boiling point much faster than would be explained by merely diluting the water.

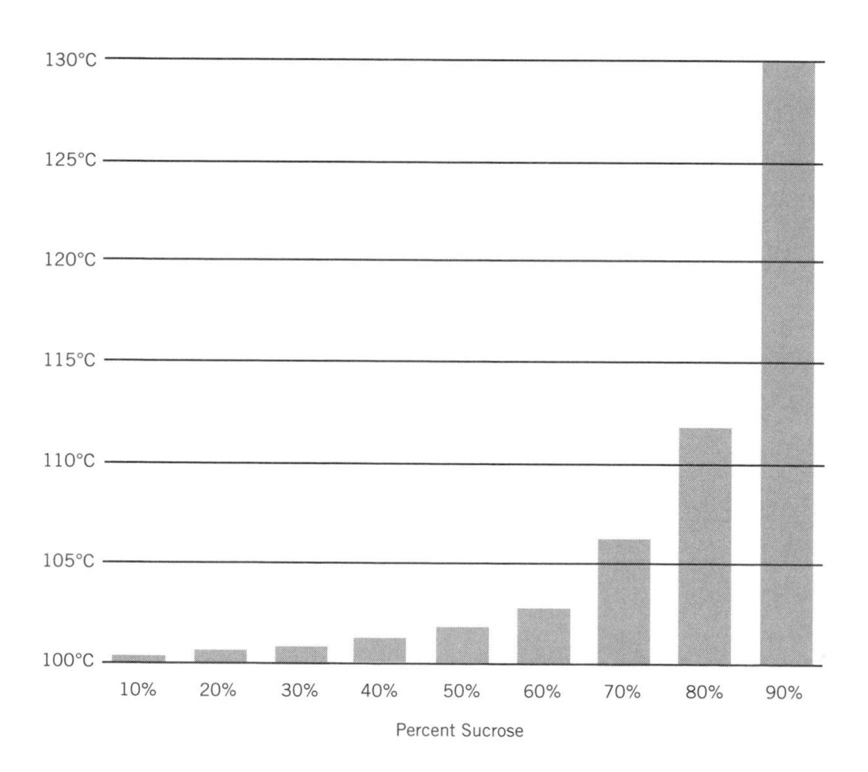

As the concentration of sucrose in the water increases, the amount of increase in the boiling point goes up even faster, unlike the linear graph that adding salt would produce. Some of the sugar molecules will break into glucose and fructose molecules. Fructose melts at 217°F (103°C), and some of the other sugars are going to be dissolved in liquid fructose, and so is some of the water. Things start to get complicated quickly.

Pressure Cookers

In the kitchen, the boiling point of water is used as a convenient temperature modulator. Since the temperature cannot exceed boiling as long as there is water in the pot, you can let the pot boil and only worry about how long you leave the food to cook.

If you could choose a higher boiling point, you could cook food faster and still only have to keep track of the time. A pressure cooker lets you do that.

Pressure cookers have several advantages over a pot of boiling water. The food is cooked at a higher temperature, so it cooks faster. When the pressure cooker is used to steam food (instead of boiling it), vitamins and minerals are kept in the food and not destroyed by light or oxygen; the steam replaces the air in the pot, and the lid keeps the light out. Also, volatile flavor molecules are not lost into the air.

Some effects of cooking are due to temperature alone, and some are due to elevating the temperature for a length of time. Physical effects tend to be in the former category, such as softening tough connective tissue in meat, turning it into gelatin, or swelling the starch grains in vegetables to create a soft gel. Allowing more time in the pot favors the second category, where chemical reactions change the food, often destroying nutrients or producing bad flavors and odors. Quickly cooking food in a pressure cooker can get the benefits of temperature without giving the food time to lose flavor and nutrition.

Some chemical changes produce flavors we like. The main one that comes to mind is the Maillard reaction. But lucky for the cook, meat can be seared before being cooked under pressure, gaining the best of both processes.

Another advantage of pressure cookers is sterilization. The higher heat destroys microorganisms faster than they are killed in boiling water. When preserving foods by canning, heating the cans or jars in a pressure cooker kills harmful organisms in much less time.

Vacuum in Canning Jars

Low pressures are also available in the kitchen. When you put food in a jar and cook it, water turns to steam in the jar, displacing the air. When you screw the top onto the jar while it is still hot, the steam is still there, keeping the air out of the jar. But as the jar cools, the steam condenses back into water. This leaves a vacuum in the space where the steam was. The pressure of the air outside has not changed, and without steam pressure inside to hold it back, air pressure bows the metal lid down into the jar.

The lid is formed with a convex ridge so that as it is bent down into the jar, it snaps into a concave form. When you open the jar, air gets in and the lid suddenly snaps back into its convex shape. The popping sound it makes lets you know that there was a vacuum inside the jar and that bacteria have not been at work making gases that fill the vacuum.

Lowering the Freezing Point

Adding a solute to water raises the boiling point, but it also lowers the freezing point, and for the same reason. The freezing point is the temperature at which as many water molecules are attaching to the ice (freezing) as there are water molecules leaving the ice (melting). If the temperature is above the freezing point, more water molecules will leave the ice than are joining it, creating a puddle. If the temperature is below the freezing point, more water molecules will stick to the ice than are leaving it, creating an ice cube.

Adding a solute such as salt or sugar changes this equilibrium. The solute does not stick to the ice like a water molecule does. It merely gets in the way, taking the place of a water molecule that might have hit the ice and stuck. However, it does not interfere with the water molecules that leave the ice. As a result, more molecules leave the ice than freeze onto it. In order to restore the balance, you need to lower the temperature until the same number of molecules are sticking as are leaving.

Making Ice Cream

Salt is used to melt the ice on roads, although sugar or some other solute would also work. Salt is used because it is cheap. But if the temperature falls below -6°F (-21°C), salt will no longer melt the ice, because it would require dissolving more salt than the water can hold—the water would be saturated with salt. The salt would just crystallize back out of the water, leaving a mixture of water, ice, and salt.

But melting ice has another useful property besides making highways less slippery: it makes things cold. It takes energy to break the bonds that hold water molecules together in ice. If you put an ice cube into room-temperature water, the energy in the fast-moving molecules in the water will get used up making chemical bonds as molecules freeze onto the ice. More molecules will leave the ice than freeze onto it, but the molecules leaving the ice will not have a lot of energy and will be moving slower, and thus they will be colder.

The cooling effect continues until the ice and the water reach their equilibrium temperature. For pure water, this is 32°F (0°C). But if you add salt or sugar, the cooling effect will continue until the temperature drops to the new equilibrium temperature. If you add a lot of salt, you will reach the point at which no more will dissolve, and the equilibrium temperature will reach –6°F (–21°C).

This is why you can freeze ice cream by putting it into a mixture of ice and salt.

INDEX